T0331018

Steel-Reinforced Concrete Structures

Steel-Reinforced Concrete Structures: Assessment and Repair of Corrosion, Third Edition examines the corrosion of reinforced concrete from a practical point of view, highlights protective design and repair procedures, and presents ongoing maintenance protocols. Updated throughout, this new edition adds additional information on concrete repair and reviews new examples of the effects of corrosion on both prestressed and reinforced concrete structures. It also examines economic analysis procedures and the probability of structural failures to define structural risk assessment and covers precautions and recommendations for protecting reinforced concrete structures from corrosion based on the latest codes and specifications.

Features:

- Updated throughout and adds all new information on advanced testing and repair techniques.
- Discusses the theoretical and practical methods of performing structural assessments.
- Explains precautions for design and construction that reduce the risk of structural corrosion.
- Covers traditional and advanced techniques for repair and how to choose the best methods.
- Utilizes the newest building codes, specifications, and standards regarding construction and corrosion.

Steel-Reinforced Concrete Structures

Assessment and Repair of Corrosion

Third Edition

Mohamed Abdallah El-Reedy, Ph.D.

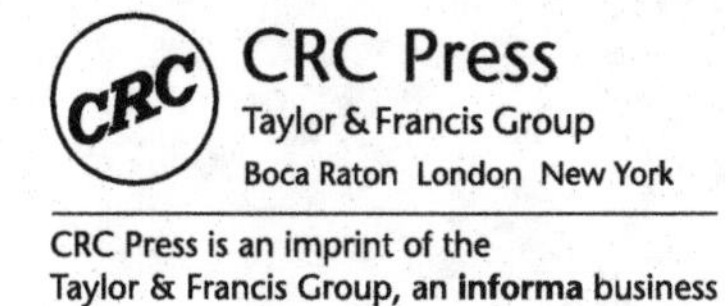

CRC Press is an imprint of the
Taylor & Francis Group, an informa business

Designed cover image: Shutterstock

Third edition published 2024
by CRC Press
6000 Broken Sound Parkway NW, Suite 300, Boca Raton, FL 33487-2742

and by CRC Press
4 Park Square, Milton Park, Abingdon, Oxon, OX14 4RN

CRC Press is an imprint of Taylor & Francis Group, LLC

First edition published by CRC Press 2007
Second edition published by CRC Press 2018

Library of Congress Cataloging-in-Publication Data
Names: El-Reedy, Mohamed A. (Mohamed Abdallah) author.
Title: Steel-reinforced concrete structures : assessment and repair of corrosion / Mohamed A. El-Reedy, Ph.D.
Description: Third edition. | Boca Raton : CRC Press, 2024. | Includes bibliographical references and index.
Identifiers: LCCN 2023009629 (print) | LCCN 2023009630 (ebook) | ISBN 9781032525310 (hbk) | ISBN 9781032525327 (pbk) | ISBN 9781003407058 (ebk)
Subjects: LCSH: Steel, Structural—Corrosion. | Reinforced concrete—Corrosion. | Reinforced concrete construction.
Classification: LCC TA445.5.E47 2024 (print) | LCC TA445.5 (ebook) | DDC 620.1/37—dc23/eng/20230302
LC record available at https://lccn.loc.gov/2023009629
LC ebook record available at https://lccn.loc.gov/2023009630

ISBN: 978-1-032-52531-0 (hbk)
ISBN: 978-1-032-52532-7 (pbk)
ISBN: 978-1-003-40705-8 (ebk)

DOI: 10.1201/9781003407058

Typeset in Times
by codeMantra

Dedication

This book is dedicated to the spirits of my mother and my father, and to my wife and my son Hisham and my daughters, Maey and Mayar

Contents

Preface

Deterioration of reinforced concrete structures has been one of the major challenges facing civil engineers over the past 30 years. Concrete structures were famous for being maintenance-free. Over time, however, it was found that nothing is free and that it was necessary to increase spending if one did not know how to deal with deterioration of concrete structures. The major factor causing concrete deterioration is corrosion of steel reinforcements, so it is very important to understand this issue thoroughly.

This book aims to be a handbook for reinforced concrete structure maintenance to be used as a guide for junior and senior engineers working in design, construction, repair, and maintenance. Moreover, this book has been written in a way that is easy for civil engineers to understand—without any complicated chemical reactions—and to help researchers with a very strong theoretical background in the field.

This book consists of three main approaches. First, Chapters 1–3 describe the corrosion phenomenon of steel in concrete, the effect of concrete properties on corrosion, and the precautions taken in construction to control corrosion. The second approach concerns the evaluation of concrete structures and different methods to protect steel reinforcement from corrosion. The last two chapters describe the most traditional and advanced repair techniques. Moreover, Chapter 9 focuses on an advanced maintenance plan philosophy as a risk-based maintenance for reinforced concrete structures.

As the first edition was successfully distributed among different industries and is considered to be a reference for many research studies and theses, for the second edition, I added the most recent research studies and development activities that were carried out in the area of corrosion of steel bars in the last 10 years and the most recent techniques for the protection and repair of concrete structures that deteriorate due to corrosion of steel bars.

As there is a continuous research and development in this area, in addition to updated code and standard for assessment this brown field. On the other hand, as per the recent earthquake in February 2023 and its damage effect in Turkia and Syria buildings, the main question how the aging structure with corroded steel bars can withstand the earthquake load. Therefore in this third edition, I try to present the best on class methodology of assessment, protection, and repair concrete structures due to corrosion.

As the main objective of this book is to provide a practical solution with a strong theoretical background, new techniques in inspection and repair have been updated to match the current market situation.

This book is a practical guide to the art of concrete structure assessment, repair, and maintenance. It includes case histories from all over the world to assist the reader in appreciating the widespread applications and range of advanced repair techniques.

Mohamed Abdallah El-Reedy, PhD
Cairo, Egypt
Email: elreedyma@gmail.com
www.elreedyman.com

About the Author

Dr. Mohamed Abdallah El-Reedy's background is in structural engineering. His main area of research is the reliability of concrete and steel structures. He has provided consultation for different oil and gas industries in Egypt and to international companies, such as the International Egyptian Oil Company (IEOC) and British Petroleum (BP). Moreover, he provides different concrete and steel structure design packages for residential buildings, warehouses, and telecommunication towers. He has participated in liquefied natural gas (LNG) and natural gas liquid (NGL) projects with international engineering firms. Currently, he is responsible for reliability, inspection, and maintenance strategies for onshore concrete structures and offshore steel structure platforms. He has performed these tasks for hundreds of structures in the Gulf of Suez in the Red Sea. He has consulted with and trained executives at many organizations, including the Arabian American Oil Company, BP, Apache, shell, Abu Dhabi Marine Operating Company (ADMA), the Abu Dhabi National Oil Company, King Saudi's Interior Ministry, Qatar Telecom, the Egyptian General Petroleum Corporation, Saudi Arabia Basic Industries Corporation (SABIC), the Kuwait Petroleum Corporation, Qatar Petrochemical Company (QAPCO), Petroleum Development Oman (PDO), and PETRONAS Malaysian oil company. He has taught technical courses on repair and maintenance of reinforced concrete structures and steel structures worldwide, especially in the Middle and Far East. Dr. El-Reedy has written numerous publications and presented many papers at local and international conferences sponsored by the American Society of Mechanical Engineers, the American Concrete Institute, the American Society for Testing and Materials, and the American Petroleum Institute. He has published many research papers in international technical journals and has authored ten books on total quality management, quality control and quality assurance, economic management for engineering projects, and repair and protection of reinforced concrete structures, advanced concrete materials, onshore structure design in petroleum facilities, and concrete structure reliability. He received his bachelor's degree from Cairo University in 1990, his master's degree in 1995, and his PhD from Cairo University in 2000.

1 Introduction

The risks of corrosion and its impact on the national income of any country are explained in Chapters 2 and 3. Corrosion of steel bars is often indicated by brown spots on the concrete surface; in some cases, cracks parallel to the steel bars will be seen, and, in the worst case, the steel bars can be seen directly because the concrete cover has failed. Typically, a structural engineer who has knowledge of the theoretical and practical methods of performing structural assessment is requested to assess the existing building. This will be discussed in Chapter 4. In Chapter 5, different codes and standards that deal with and provide corrosion precautions for design and construction to reduce the risk of corrosion on a structure are explored. Corrosion control is illustrated in Chapter 6, and the standard steps of repair, traditional and advanced repair techniques, and ways to choose the best method of repair are covered in Chapter 7. There are many research and development techniques to protect steel reinforcement bars; all the practical methods available on the market are illustrated in Chapter 8. The new approach of structure integrity management system will be presented in Chapter 9. In addition to that, in this chapter, all methods that will assist in implementing the maintenance plan from an economic point of view, and the economic methods used to choose among steel protection alternatives will be discussed.

The civilization of any country is measured by the advances in the techniques used in the construction of its buildings. Concrete is the main element used in construction materials, and the quality of concrete being used for construction in any country is an indication of that country's progress in engineering thinking. Ancient Egypt used concrete in its buildings and temples; Egyptians used crushed stone as an aggregate and clay as an adhesive. The Greeks used concrete in their buildings and called it *Santorin Tofa* (El-Arian and Atta 1974); the Romans used a material called *pozzolana* that resembled concrete. After that, the formula for concrete was lost for many centuries. In the eighteenth century, a number of famous scientists worked to create a formula for concrete:

- John Smeaton used concrete to construct the Eddystone Lighthouse.
- Joseph Parker researched stone and its uses in concrete.
- Edgar researched using cement made from limestone and clay.
- Louis Vicat also tried to develop cement from limestone and clay.
- Joseph Aspdin completed his research by developing portland cement.

At the end of the nineteenth century and in the beginning of the twentieth century, the architectural shapes of buildings underwent major changes as architectural engineers and builders changed their points of view. Previously, ideas from the European Renaissance of using columns and arches to move toward the functional use of

DOI: 10.1201/9781003407058-1

buildings have been implemented. Concrete has been found to be the best choice for functional use and to achieve architectural dreams using an economic approach.

Currently, reinforced concrete is considered to be the most popular and important material in the construction industry. It is used for buildings and for different types of civil engineering projects, such as tunnels, bridges, airports, and drainage and hydraulic projects. Research has been focused on increasing the performance of concrete to match a wide variety of applications.

Reinforced concrete is considered to be inexpensive compared to other building materials; thus, it has been used for high-rise buildings for megaprojects and also for smaller projects, such as one-story buildings; all these projects are handled by contractors and engineers with different capabilities. Therefore, sometimes, the reinforced concrete does not perform correctly according to specifications; in some cases, there is a difference between the quality required by the standard or code and that found on the site. These errors usually arise due to a lack of worker, foreman, or engineer competence. Other construction errors arise due to a lack of quality-control procedures, materials received at the site that are incompatible with the standard, improper selection of materials, or a concrete mix that does not fit the surrounding environmental conditions.

The main challenge that is faced by engineers working in the concrete industry today is the corrosion of steel reinforcement bars. When concrete was first used, concrete structures were traditionally treated as structures that were maintenance free during their lifetime. For many years, it has been obvious that this viewpoint is false, although concrete is indeed a durable material. Our dreams have broken on the rock of truth as we have found that a number of concrete structures have undergone major deterioration because of different factors and that 80% of the deterioration of reinforced concrete structures is due to the corrosion of the steel-reinforced bars. Based on this fact, much research and many development techniques have focused on providing materials and new methodology to protect steel reinforcement bars from corrosion.

Because severe climatic conditions are the major contributing factor of corrosion in reinforcement steel bars, all specifications and codes recommend some precautions to be taken in concrete mix design, material selection, concrete cover thickness, and others in order to maintain the durability of concrete during its lifetime and to prevent corrosion of the steel bars based on different environmental conditions. Corrosion of steel is the enemy of any country's investment in real estate; some structures have been completely destroyed due to corrosion and depleted steel bars have a direct impact on the safety of a structure.

Billions of dollars have been spent on the corrosion problem worldwide. For example, the United States has spent billions of dollars to repair bridge decks. In North Africa, near the coast, and in the Middle East in the Arabian Gulf area, some buildings have been completely destroyed due to deterioration of the structures as a result of corrosion in the steel reinforcement bars.

It is important to know the causes of corrosion, the concrete parameters, and the environmental conditions surrounding the structure that trigger corrosion and affect the corrosion rate. For instance, it is acceptable in European and Middle Eastern countries to use seawater in the concrete mix. In addition, in 1960, calcium chloride was used as an additive to assist in concrete-setting acceleration, and seawater was used in concrete mix until 1970. After that, much damage to the concrete structure was found over a 20-year period—especially in Middle Eastern countries where, because potable water is very expensive, they used seawater in most of the concrete mixes for different structures. This period had a very bad impact from an economic point of view because the investment for the construction of concrete structures was destroyed. From the aformentioned facts, the risk of corrosion and its impact on the national income of any country become clear, and therefore, an understanding of corrosion characteristics and causes of corrosion, which are discussed, respectively, in Chapters 2 and 3, becomes important.

Usually, corrosion of steel bars is indicated by brown spots on the concrete surface; in some cases, cracks parallel to the steel bars will be seen and, in the worst case, the steel bars can be seen directly because the concrete cover has failed. When corrosion occurs, its major impact on structure safety will be the danger of the falling of concrete cover, as is the case when the structure loses its strength due to reduction in the concrete cross-section dimension and also reduction in the cross-section of the steel bars due to corrosion. The spalling of concrete cover happens because the corroded steel bars increase in volume and exert high stress on the concrete cover; this causes cracking and subsequent falling of the cover.

Usually, a structural engineer who has knowledge of the theoretical and practical methods of performing structural assessment is requested to assess the existing building. This will be discussed in Chapter 4. All the costs for repair and reconstruction have a higher impact on a country's economy in general because the money paid for repair and reconstruction could have gone into other investments.

Chapter 5 discusses different codes and standards that deal with corrosion and provides precautions to be taken in design and construction to reduce the risk of corrosion in a structure. Corrosion control is illustrated in Chapter 6. Chapter 7 provides all the steps for repair based on standards, traditional and advanced techniques of repair, and ways to choose the best method of repair. There are many research and development techniques to protect steel reinforcement bars; all the practical methods available on the market are illustrated in Chapter 8.

During the last 30 years, some researchers have studied the reliability of reinforced concrete structures and have provided different techniques, such as qualitative and quantitative risk assessments. In addition, the maintenance strategy and plan have been reviewed in different studies, based on risk-based maintenance. Our aim is to present the new approach of asset integrity management that is used widely these days. The objective of this management tool is to use an optimal methodology that reduces maintenance costs to a minimum and maintains the structure within a reliability limit that enables functional use to user satisfaction. Methods that will assist in implementing the maintenance plan from an economic point of view, as well as the economic method of choice among steel protection alternatives, are discussed in Chapter 9.

REFERENCE

El-Arian, A. A. and A. M. Atta. 1974. *Concrete Technology*. Giza, Egypt: World Book.

2 Corrosion of Steel in Concrete

2.1 INTRODUCTION

In the past, most of the design studies in the literature and research in reinforced concrete assumed that the durability of reinforced concrete structures could be taken for granted. However, many reinforced concrete structures are exposed during their lifetimes to environmental stress (e.g., corrosion and extensive aggregation reactions), which attacks the concrete or steel reinforcement (Cady and Weyers 1984; Kilareski 1980; Mori and Ellingwood 1994; Takewaka and Matsumoto 1988; Thoft-Christensen 1995). Corrosion of reinforced steel bars has been considered to have been the most famous problem facing structural engineers in the last decade. In order to avoid the corrosion of the steel bars in a concrete structure, we need to understand the corrosion phenomenon so that we can provide different engineering solutions to this problem.

Because corrosion of steel bars in reinforced concrete structures is very expensive, it is mainly considered to be an economic rather than an engineering problem. In the United States, the cost of repair due to the corrosion of steel bars in buildings and bridges is around $150 million per year. In countries like the United States, Canada, and Europe, when ice accumulates on bridges during cold weather, salt is usually used to melt the ice. However, as we will discuss, salt contains sodium chloride, which is the main source of corrosion of steel in these countries, so the cost of maintaining and repairing bridge decks is very high. A transportation research center report in 1991 indicated that the cost of bridge deck repair ranged from $50 to $200 million per year; for some parts of bridges, repair costs were around $100 million per year. Repairs in multifloor parking garages cost $50–$150 million per year. In England and Wales, bridge repair due to corrosion cost £616.5 million in 1989. Considering that these two countries have only 10% of the United Kingdom's bridges, imagine the impact of corrosion on the countries' economies!

However, in the Middle East, especially in the gulf area, countries that are the main producers of oil and gas worldwide have very strong economies. As a result, it is possible for them to make huge investments in construction projects, which are the main part of the economic growth of any developing country. The bad news is that all these countries have harsh environmental conditions due to high temperatures, which can reach 55°C–60°C. In addition, because most of these countries are located near the Arabian Gulf, they have a high relative humidity and the groundwater has a high percentage of salt.

DOI: 10.1201/9781003407058-2

All the aforementioned factors increase the probability of corrosion of steel bars. From the quality control point of view, a higher temperature prevents a very good curing process. Thus, deterioration of concrete structures in these countries is accelerated and causes their buildings to have short lifetimes—which is what occurred especially for buildings constructed in the period between 1970 and 1980 (before the practice of using additives in concrete). Therefore, in the Middle East, the amount of money spent due to corrosion of steel bars is very high; thus, the development of new methods and techniques to protect these bars is considered a challenge to engineers.

The corrosion process occurs slowly and propagates with time, so the deterioration rate varies. Corrosion of steel bars affects a structure's safety and depends on the surrounding environmental conditions that mainly affect the corrosion rate, the location of the member in the building, and the type of the member.

When should inspection and repair be performed? This is a big question that needs an accurate answer because reducing the time between inspections will increase the cost but will maintain the safety of the structure. However, increasing the time between inspections can decrease the cost but will affect the safety of the structure. Many accidents have happened due to noncompliance with the above; for example, a large part of a bridge in New York fell, causing a motorcycle rider to lose control of his vehicle and die, thus making it obvious that carelessness with regard to the corrosion process may also result in loss of life.

Figure 2.1 shows a clear example of the effect of corrosion on concrete structure deterioration. It shows cracks and the concrete cover and parts of the plaster that are falling. This reinforced concrete beam is in a villa near the coast of the Mediterranean Sea.

In 1960, in Europe and the Middle East, it was acceptable to use seawater in the concrete mixture as well as to use calcium chloride additives as a concrete-setting accelerator; these practices continued until 1970. In the Middle East, pure water is very expensive, so they used seawater extensively; after 20 years, they found severe deterioration in the reinforced concrete structures, causing many problems from an economic point of view (i.e., construction investments).

FIGURE 2.1 Deterioration in a reinforced concrete beam.

From the preceding discussion, it is obvious that the effect of corrosion not only creates a serious problem for structural engineers but also affects the national incomes of countries. Therefore, engineers must have an in-depth understanding of the nature of the corrosion process and the reasons for its occurrence. Understanding the problem is the first step in solving the problem.

In this chapter, the main principles of corrosion will be discussed—especially corrosion of the steel reinforcements in concrete. How and why corrosion occurs will be discussed. There are many ways to illustrate the corrosion process in concrete; we will illustrate it by using first principles, which will match with engineering studies; we will try to stay away from chemical equations. Understanding the corrosion process and how it occurs will be the first step to knowing how to protect steel bars from corrosion and to differentiate between the different protection methods that will be illustrated later.

If we place a nail in dry air or immerse it in water, it will not corrode. However, if it is immersed in water and then placed in air, it will corrode. This happens for any type of normal or high-strength steel. In the case of steel embedded in concrete, as the concrete is a porous material containing water in the voids due to the process of curing or because of rainy weather or any weather with a high relative humidity, the concrete will contain moisture, which is a common cause of corrosion. The good news is that it is not necessary that steel bars embedded in concrete will corrode because concrete is alkaline in nature and alkalinity is the opposite of acidity. As a result of its alkalinity due to a high concentration of the oxides of calcium, sodium, and magnesium inside the microvoids of the concrete, it can protect the steel from corrosion.

The oxides of calcium, sodium, and magnesium produce hydroxides that have a high alkalinity in water (pH 12–13). pH is the measure of acidity and alkalinity and is based on the percentage of hydrogen ion concentration with respect to hydroxide ions; the maximum acid concentration occurs when $pH = 1$ and the highest alkaline concentration is present at $pH = 14$. The alkalinity produces a passive layer on the steel reinforcement surface and consists of oxides and hydroxides of iron and partly cement. This layer is dense and prevents the occurrence of corrosion. This layer does not remain for long. Two factors always contribute to the breaking of the layer: the carbonation and the permeability of chlorides to the steel reinforcement. This will be illustrated in detail in Chapter 3; here, the general corrosion process will be illustrated without considering the effects of any external factors on triggering the corrosion process.

2.2 THE CORROSION PROCESS

After the passive layer is broken down, rust will appear instantly on the surface of the steel bar. The chemical reactions are the same in the cases of carbonation or chloride attack. When the corrosion of the reinforced steel bars in concrete occurs, they melt in the void that contains water. The electrons accumulate according to the following equation, which presents the *anodic reaction*:

$$Fe \rightarrow Fe^{2+} + 2e^{-} \tag{2.1}$$

If the electrons are accumulated on the other part of the steel reinforcement but cannot accumulate in huge numbers in the same location, there is another reaction that uses the electrodes with oxygen and water—the *cathodic reaction*. The equation for the cathodic reaction is as follows:

$$2e^- + H_2O + \frac{1}{2}O_2 \rightarrow 2OH^- \tag{2.2}$$

From this equation, it is found that the formation of OH^- occurs due to the cathodic reaction. The hydroxide ions increase the alkalinity and reduce the effect of carbonates or chlorides slightly. From this equation, it is important to know that water and oxygen are the main reasons for the occurrence of corrosion process.

As shown in the preceding equations and Figure 2.2, the anodic and cathodic reactions are the first steps in the process of corrosion as the hydroxide ions (OH^-) will react with ferrous irons (Fe^{2+}) as a result of the chemical equation (Equation 2.1). This reaction will produce ferrous hydroxide, which will react with oxygen and water again and produce ferric hydroxide. This chemical reaction is shown graphically in Figure 2.2.

$$Fe^{2+} + 2OH^- \rightarrow Fe(OH)_2 \tag{2.3}$$

$$4Fe(OH_2) + O_2 + 2H_2O \rightarrow 4Fe(OH)_3 \tag{2.4}$$

$$2Fe(OH)_3 \rightarrow Fe_2O_3 \cdot H_2O + 2H_2O \tag{2.5}$$

The preceding chemical reactions show the transformation of steel from ferrous hydroxides ($Fe(OH)_2$), which will react with oxygen and water to produce ferric hydroxides ($Fe(OH)_3$), and the last component, which is hydrated ferric oxide (rust); its chemical term is $Fe_2O_3 \cdot H_2O$.

Ferric hydroxide has a greater effect on concrete deterioration and spalling of the concrete cover as its volume will increase the volume of the original steel bars by about two times or more. When iron forms hydrated ferric oxides in the presence of water, its volume increases further to reach about ten times its original volume

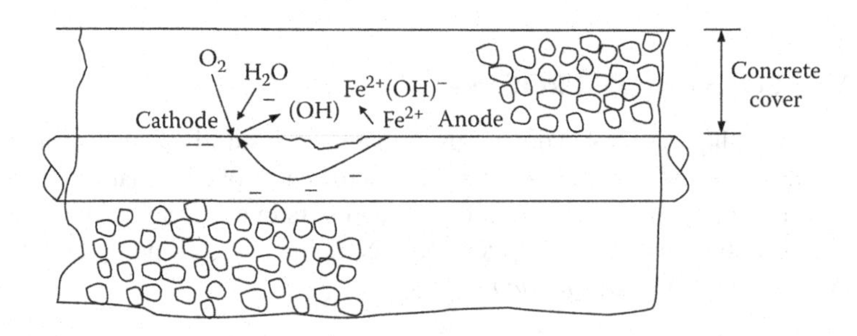

FIGURE 2.2 Corrosion process on a steel reinforcement surface.

and it becomes soft. At this stage, cracks start forming on concrete and the process continues until the concrete cover falls; rust, with its brown color, can clearly be seen on the steel bar.

2.3 BLACK CORROSION

This type of corrosion occurs when there is a large distance between the anode and cathode and also if oxygen is not available. This usually occurs in cases of buildings immersed in water or when a protective layer blocks the availability of oxygen. It is called black corrosion because, when it cracks, the bars will have a black or green color. This type of corrosion is critical because it does not provide any warning of cracks or falling concrete cover when corrosion occurs. This type of corrosion occurs usually in offshore structures with parts submerged in water or in a concrete structure partially covered by a waterproof membrane or coating.

2.4 PIT FORMATION

Corrosion in steel bars starts by forming a small pit. The number of pits increases with time and then the pits combine to form a uniform corrosion on the surface of the steel bars. This is obvious in the case of a steel reinforcement exposed to carbonation or chloride effects, as shown in Figure 2.3. The pit formation is shown in Figure 2.4.

Many chemical reactions describe the formation of pits, and, in some cases, these equations are complicated. But the general principle of pit corrosion is very simple, especially in cases of chloride attacks. In some locations on the steel reinforcement, voids in the cement mortar are present around the steel reinforcement or sulfide is present inside steel bars so that the passive layer is more vulnerable to chloride attack; an electrochemical potential difference attracts chloride ions. Corrosion is initiated and acids are formed as hydrogen from the supplied MnS inclusion in steel

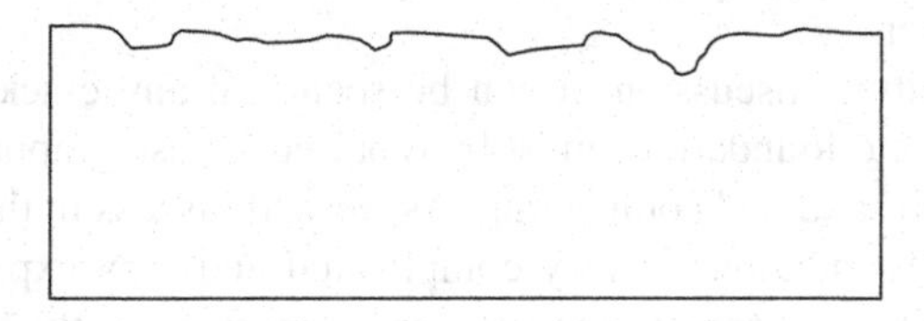

FIGURE 2.3 Uniform corrosion.

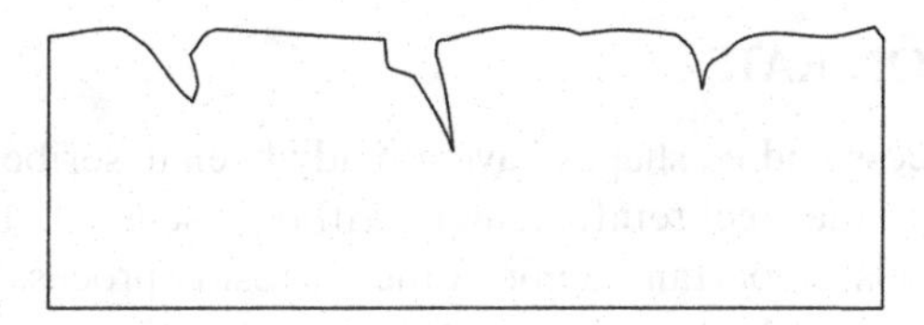

FIGURE 2.4 Pitting corrosion.

and HCl from the chloride ions, if they are present. The following chemical reactions are simple representations of this process:

$$Fe^{2+} + H_2O\,FeOH^{+} + H^{+} \tag{2.6}$$

$$MnS + 2H^{+}H_2S + Mn^{2+} \tag{2.7}$$

Rust may form over the pit, concentrating the acid (H^+) and blocking oxygen so that the iron stays in solution, preventing the formation of a protective oxide layer and accelerating corrosion. We will return to the subject of pitting corrosion later. It is related to the problems of reinforcement coating and to the black rust phenomenon that will be discussed.

This state of corrosion is characterized by galvanic action between a relatively large area of passive steel acting as the cathode and a small anodic pit where the local environment inside the pits has a high chloride concentration and decreased pH value. For pitting to be sustained, it is necessary that a reasonable amount of oxygen should be available to cause polarization of the anode. The average corrosion potential of steel reinforcement has a pitting that is likely to vary between that of the passive state and that of the anodic pitting areas—typically in the range of −200 to −500 mV.

2.5 BACTERIAL CORROSION

Bacteria are another cause of corrosion. Because bacteria exist in the soil, the foundation is considered the main element exposed to this type of corrosion. These bacteria will convert sulfur and sulfides to sulfuric acid. The acid will attack the steel and then cause the initiation of the corrosion process. Other bacteria that attack the sulfide exist in the steel reinforcement FeS due to reactions. This type of corrosion is often associated with a smell of hydrogen sulfide (rotten eggs) and smooth pitting with a black corrosion product when steel bars are exposed to soil saturated with water.

From the preceding discussion, it can be seen that any cracks or honeycombs in reinforced concrete foundations must be repaired by using appropriate materials before refilling with sand and compacting. As we will discuss in the following chapters, the repair of foundations is very complicated and very expensive. Therefore, avoiding these defects in construction is very important to the maintenance of a structure throughout its lifetime.

2.6 CORROSION RATE

The corrosion process and its shapes have already been described. In this section, the corrosion rate of the steel reinforcement will be discussed. The corrosion rate is considered the most important factor in the corrosion process from a structural safety perspective and in the preparation of the maintenance program for the structure. This factor is considered an economic factor for structural life. When the corrosion rate is very high, the probability of structure failure will increase rapidly and structural safety will reduce rapidly. The corrosion rate depends on different factors, so if we can control these factors, the corrosion rate will be low. The corrosion will occur, but it will not cause a serious problem to the structure if the rate of corrosion is low.

The main factor that affects the corrosion rate is the presence of oxygen, especially in the cathodic zone shown in the previous chemical reactions and in Figure 2.5. In the case of nonavailability of oxygen, the corrosion rate will be slow and different methods are used to prevent the propagation of the oxygen inside the concrete—for example, to take care of the concrete compaction in order to obtain a dense concrete cover so that propagation of the oxygen will be very slow or is prevented. Practically speaking, this is an ideal case that cannot happen but that we will try to reach. When we prevent the propagation of oxygen inside the concrete, the oxygen in the steel bars will be less. The difference in volts between anodic and cathodic zones will also be less—an effect on corrosion called "polarization." The Evans diagram in Figure 2.6 shows the polarization curves separately for anodic and cathodic reactions intersecting at a point (P), where the mean anodic and cathodic current densities are equal and represent the corrosion rate in terms of a mean corrosion current density, I_{corr}. The electrode potential of the couple at this point is termed the corrosion potential, E_{corr}.

The second most important factor affecting the corrosion rate is the moving of the ions inside the concrete voids around the steel reinforcement. If the speed of the moving ions is very low or is prevented, the corrosion rate will also be very slow or, in the ideal case, is prevented. This case may occur when the concrete around the steel bars has a high resistance to electrical conductivity between the anode and the cathode.

The measurement of electrical resistivity to the concrete surrounding the steel reinforcement can give us an idea of the corrosion rate and the chemical reaction rate. There are many ways to measure corrosion rate by measuring the electrical resistivity, as will be described in detail in Chapter 4.

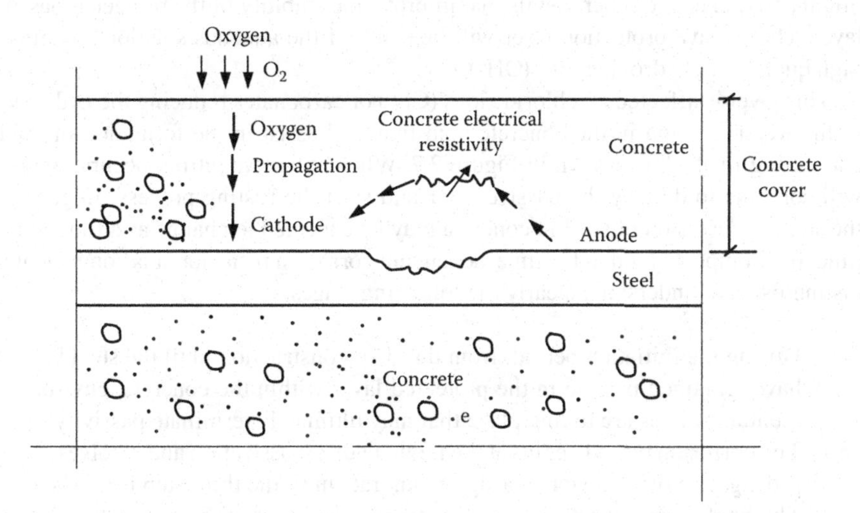

FIGURE 2.5 Factors affecting corrosion rate.

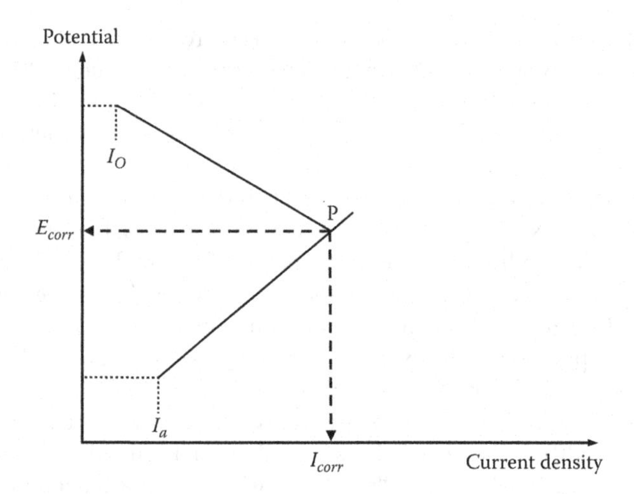

FIGURE 2.6 Evans diagram for a simple corrosion couple.

There are different types of metals that can be used to control the corrosion rate. One among them is steel, which reacts with oxygen and forms a layer of thickness 0.01 μm of oxides of the same metal at the surface; this layer is stable and protects it from breaking under any circumstances. The metal is considered protected when it is inside the liquid solution and reduces the melting of ions in the liquid solution surrounding the steel bars so that the corrosion rate will be very low if this passive layer exists. The corrosion rate will be neglected and considered to be zero (Figure 2.5).

In this case, this layer will be the passive protection layer for steel reinforcement. The protected passive layer is first responsible for protection from corrosion, and this is obvious in the case of stainless steel. The steel in this case consists of some chromium and nickel and other metals that improve the stability of the protected passive layer. The passive protection layer will be stable if the aqueous solution contains a high quantity of hydroxide ions (OH^-).

This layer is affected by chloride ions (CL^-) or carbonates reducing the hydroxide in the water solution in the concrete void that will assist in the formation of voids and pitting in steel, as shown in Figure 2.7. When extensive pitting occurs, the pits will combine to destroy the passive layer and start the rusting process. In general, the state of corrosion of steel in concrete may be expected to change as a function of time. In attempts to model this time-dependent corrosion behavior, it is convenient to distinguish and understand clearly the following stages:

- During the initiation period, from day 1 of construction until the steel bars have remained passive in the protected layer within the concrete, environmental changes are taking place that may ultimately terminate passivity.
- The corrosion period begins at the moment of depassivation and involves the propagation of corrosion at a significant rate until the third step is reached.
- The final stage is reached when the structure is no longer considered acceptable on grounds of structural integrity, serviceability, or appearance.

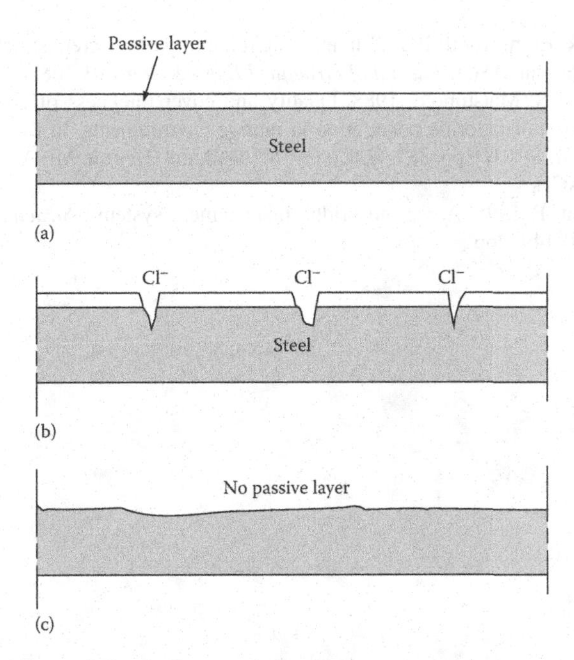

FIGURE 2.7 Effect of chlorides on demolishing the passive layer: (a) passive layer, (b) chloride attack, and (c) corrosion on passive layer.

2.7 SUMMARY

This chapter has illustrated the corrosion process, reasons for its occurrence, the accompanying chemical reactions, electron movement, and the formation of oxides. It is important to understand the shape of corrosion and its type, such as pit formation, black corrosion, and bacterially induced corrosion. Understanding anode and electron formation is important in the protection of steel bars.

Other chapters will discuss how corrosion can occur and the factors that affect it. Understanding corrosion is a very important step to obtaining engineering solutions to avoid it, as well as obtaining the best methods for protection and repair of reinforced concrete structures exposed to corrosion. You must know your enemy before you can fight him.

It is important to mention that the quality of concrete and the thickness of the concrete cover of the steel reinforcement affect the stability of the passive layer. They influence the ability of the system to exclude aggressive substances, which tend to alter the pore–water composition in ways that endanger the passivity of the embedded steel and thus induce significant corrosion.

REFERENCES

Cady, P. D. and R. E. Weyers. 1984. Deterioration rates of concrete bridge decks. *ASCE Journal of Transactions in Engineering 110(1)*:34–44.

Kilareski, W. P. 1980. Corrosion-induced deterioration of reinforced concrete—An overview. *Materials Performance 19(3)*:48–50.

Mori, Y. and B. R. Ellingwood. 1994. Maintaining reliability of concrete structures. I: Role of inspection/repair. *ASCE Journal of Structural Engineering 120*(*3*):824–845.

Takewaka, K. and S. Matsumoto. 1988. Quality and cover thickness of concrete based on the estimation of chloride penetration in marine environments. In *Concrete in Marine Environment*, SP-109, pp. 381–400, ed. V. M. Malhotra. Detroit, MI: American Concrete Institute (ACI).

Thoft-Christensen, P. 1995. Advanced bridge management systems. *Structural Engineering Review 7*(*3*):149–266.

3 Causes of Corrosion and Concrete Deterioration

3.1 INTRODUCTION

The process of steel corrosion, how it occurs, and the key concepts applicable to it are discussed in Chapter 2. This chapter discusses different kinds of mechanical corrosion of steel in concrete structures. There are two main reasons for corrosion of steel in concrete: chloride attack and carbon dioxide penetration, which is also called carbonation. These two processes cause corrosion of steel, but they do not influence corrosion of steel in concrete directly. Instead, there are other reasons, such as the presence of certain chemicals inside the concrete and voids that affect the steel. Moreover, some acids, such as sulfate, can attack concrete, cause concrete deterioration and corrosion of steel, and then break concrete's alkalinity around steel bars.

When acids attack concrete, there is an impact on steel reinforcement. The main factors that cause the corrosion in reinforced concrete structures exposed to carbonation or chlorides are discussed here, as well as the basics of corrosion and how they apply to steel in concrete, corrosion rate, and corrosion effects on spalling of concrete.

Concrete is alkaline, which is the opposite of acidic. Metals corrode in acids, whereas they are often protected from corrosion by alkalis. Concrete is alkaline because it contains microscopic pores with high concentrations of soluble calcium, sodium, and potassium oxides. These oxides form hydroxides derived from the reactions between mixing water and portland cement particles, which are highly alkaline. The measuring factor of the acidity, alkalinity, and pH is based on the fact that the concentration of hydrogen ions (acidity) times hydroxyl ions (alkalinity) is 10^{-14} mol/L in aqueous solution. A strong acid has $pH = 1$ (or less), a strong alkali has $pH = 14$ (or more), and a neutral solution has $pH = 7$. Concrete has a pH of 12–13; steel starts to rust at pH 8–9.

Concrete creates a highly alkaline condition within the pores of the hardened cement mix that surrounds the aggregate particles and the reinforcement. This alkaline condition leads to a "passive" layer on the steel surface. A passive layer is a dense, impenetrable film, which, if fully established and maintained, prevents further corrosion of steel. The layer formed on steel in concrete is probably part metal oxide/hydroxide and part minerals from the cement. A true passive layer is a very dense, thin layer of oxide that leads to a very slow rate of oxidation (corrosion). Once the passive layer breaks down, areas of corrosion start appearing on the steel surface. The chemical reactions are the same whether corrosion occurs by chloride attack or by carbonation.

DOI: 10.1201/9781003407058-3

Chloride attack and carbonation transformation are unusual in that they do not attack the integrity of the concrete. Instead, aggressive chemical species pass through the pores in the concrete and attack the steel. This is unlike normal deterioration processes due to chemical attack on concrete. Other acids and aggressive ions, such as sulfate, destroy the integrity of the concrete before the steel is affected. Most forms of chemical attacks are therefore concrete problems before they are corrosion problems. Carbon dioxide and chloride ions are very unusual in that they penetrate the concrete without significantly damaging it. Accounts of acid rain causing corrosion of steel embedded in concrete are unsubstantiated. Only carbon dioxide and chloride ions have been shown to attack the steel and not the concrete.

3.2 CARBONATION

Carbonation is the result of the chemical reaction between carbon dioxide gas in the atmosphere and the alkaline hydroxides in the concrete. Like many other gases, carbon dioxide dissolves in water to form an acid. Unlike most other acids, the carbonic acid does not attack the cement paste, but rather neutralizes the alkalis in the pore water, mainly forming calcium carbonate:

$$CO_2 + H_2O \rightarrow H_2CO_3$$

$$H_2CO_3 + Ca(OH)_2 \rightarrow CaCO_3 + 2H_2O$$

Calcium hydroxide exists in concrete and increases its alkalinity that maintains a pH level of 12–13; after carbonates attack inside the concrete and spread, they form calcium carbonate. As seen in the equation, the value of pH will be reduced to the level that causes the corrosion in the steel reinforcement, as shown in Figure 3.1.

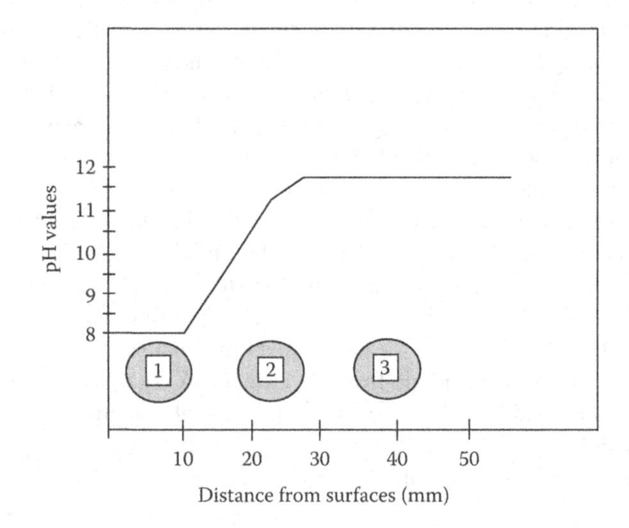

FIGURE 3.1 Relation between carbonation depth and level of pH values.

Carbonation occurs quickly when the concrete cover is not very thick. It may also occur when the concrete cover over the steel bars is thick because carbonation transformation will happen as a result of the existence of open pore voids in the concrete that assist quick propagation of CO_2 inside the concrete. Carbonation can occur when the alkalinity in voids is relatively small. This happens when the cement content is small and water-to-cement (*w/c*) ratio is high or also due to bad curing process during construction.

Carbonation moves inside the concrete according to diffusion theory: The diffusion rate is inversely proportional to the distance between the steel bars and the concrete surface, which is the concrete cover thickness:

$$\frac{dx}{dt} = \frac{D_o}{x} \tag{3.1}$$

where

x is the distance from the concrete surface
t is the time
D_o is the diffusion rate, which depends on the quality of the concrete

When the concrete is affected by carbonation, the value of alkalinity drops from pH 11–13 to pH < 8, as shown in Figure 3.1. At this level of alkalinity, the passive layer cannot protect the steel, and hence corrosion begins.

Many factors affect the ability of concrete to resist the spread of carbonation. Note that the rate of carbonation depends on the thickness of the concrete cover and on its quality in terms of the mixing ratios that achieve the highest quality. This is necessary to resist the spread of carbonation inside concrete.

When concrete is of high quality and compaction and curing are done well, it would be difficult for CO_2 to spread inside the concrete. As mentioned, if compaction is good enough to reduce the voids in concrete, it will enhance protection of concrete against carbon dioxide; this is possible by using new materials such as silica fume, which has a very fine particle size (around 0.1–0.2 μm) that will reduce the voids in the concrete. Currently, silica fumes with plasticizer are used to produce high-strength concrete, with a compressive strength of around 120 N/mm^2 and more. Moreover, in the mid-1980s, the Japanese produced concrete without compaction with a high percentage of silica fumes and a special concrete mix.

Different specifications and codes have identified an appropriate thickness for the concrete cover, depending on the surrounding environmental conditions. In addition, they have identified the *w/c* ratio in concrete mix, the reasonable cement content, the suitable compaction method, and the curing time required to avoid carbonation transformation in the concrete. Chapter 4 will explain various codes in detail.

Carbonation is the main reason for corrosion of reinforcing bars in old structures, those that have been poorly constructed, or structures containing a small proportion of cement content in the concrete mix. Carbonation rarely occurs in modern concrete bridges and new buildings, where the proportion of water to cement is low, cement content is sufficient, and compaction and concrete curing are very good. In addition, concrete cover in these structures will be sufficient to prevent the spread of carbonation to the steel reinforcement throughout the structure's lifetime.

In structures that are exposed to seawater or dissolved salt, the chlorides in concrete often penetrate until they reach the steel reinforcement bars; they cause steel corrosion that is faster than the impact of the process of carbonation. The dry and wet cycles on the surface of concrete help speed up the process of carbonation; entry of carbon dioxide gas is permitted in the dry cycle and it is dissolved in the wet cycle. This is considered a problem in some coastal cities exposed to cycles of wet and dry conditions, such as Hong Kong.

If it is found that corrosion has occurred as a result of inadequate concrete cover thickness, then the concrete around the steel reinforcement is carbonated and is causing the corrosion. When the concrete cover is not very thick, the process of carbonation transformation will speed up. This situation may lead to the process of corrosion beginning 5 years from the year of construction. When the concrete is of good quality, carbonation will not spread easily and the inadequate thickness of the concrete cover will not have a huge effect.

Carbonation can be observed and measured easily. pH indicates the proportion of phenolphthalein solution in water and alcohol, which will be sprayed or painted on the concrete surface that is expected to be affected by carbonation; phenolphthalein will change its color upon change in pH. Phenolphthalein is colorless when pH is low (carbonation). It turns pink when pH increases (concrete without carbonation). This test can be done by taking samples of concrete (usually the part of the cover that falls) so that the surface will be ready for the test or to crack part of the concrete when periodic maintenance is performed. The surface whose pH is to be measured should be clean and free from dust and other fine materials. From Figure 3.1, it can be seen that different pH values exist when carbonation is present in or is absent from concrete; this test will be explained in Chapter 5.

3.3 SPREADING OF CARBONATION INSIDE CONCRETE

The spreading of carbon dioxide inside the concrete and the rate of movement of carbonation almost follow Fick's law in circumstances in which the diffusion rate is inversely proportional to the distance from the surface, as given in the previous equation presented earlier. Since the carbonation process is considered to change the characteristics of pore voids in the concrete, reduce the base, and reduce the percentage of pH, this equation is approximate. Each of the divisions (one of the characteristics of concrete), the change in the components of concrete, and change of humidity level with depth will deviate from the values of diffusion calculated from the previous law, as given in the previous equation.

By performing integration in Equation 3.1, we will obtain the second root that assists in defining carbonation movement. This equation and some research will help obtain some of the equations that determine the relationship between the rate of carbonation spread and the quality of concrete and the environmental factors surrounding a structure. Table 3.1 summarizes some of those equations and explains the transactions included in them.

TABLE 3.1
Equations to Calculate Carbonation Depth

Equations	Coefficients
$d = A(t)^{n}$	d = carbonation depth t = time in years A = diffusion factor n = exponent (approximately 0.5)
$d = ABC\,t^{0.5}$	A = 1.0 for external exposure B = 0.07–1.0 depending on surface finish $C = R(w/c - 0.25)/(0.3(1.15 + 3w/c))0.5$ For water-to-cement ratio $(w/c) \geq 0.6$ $C = 0.37R\,(4.6w/c - 1.7)$ for $w/c < 0.6$ R = coefficient of neutralization, a function of mix design and additives
$d = A(Bw/c - C)t^{0.5}$	A is a function of curing B and C are a function of fly ash used
$d = 0.43(w/c - 0.4)(12(t - 1))^{0.5} + 0.1$	28-day cured
$d = 0.53(w/c - 0.3)(12t)^{0.5} + 0.2$	Uncured
$d = (2.6(w/c - 0.3)^{2} + 0.16)t^{0.5}$	Sheltered
$d = ((w/c - 0.3)^{2} + 0.07)t^{0.5}$	Unsheltered
$d = 10.3e^{-0.123f28}$ at 3 years	Unsheltered fX = strength at day X
$d = 3.4e^{-0.34f28}$ at 3 years	Sheltered
$d = 680(f28 + 25)^{-1.5} - 0.6$ at 2 years	
$d = A + B/f28^{0.5} + c/(\text{CaO} - 46)^{0.5}$	CaO is alkali content expressed as CaO
$d = (0.508/f35^{0.5} - 0.047)(365t)^{0.5}$	
$d = 0.846(10w/c/(10f7)^{0.5} - 0.193 - 0.076w/c)(12t)^{0.5} - 0.95$	
$d = A(T - t_i)t^{0.75}\,(C_1/C_2)^{0.5}$	t_i = induction time T = temperature in kelvins $C_1 = CO_2$ concentration $C_2 = CO_2$ bound by concrete

Source: Parrott, L.J., *Mater. Protect.*, *6*, 19, 1987.

However, these are only approximations, because the carbonation process modifies the concrete pore structure as it proceeds. Cracks change based on concrete composition, and moisture levels with depth will also lead to deviation from a perfect diffusion equation. Integration of Equation 3.2 gives a square root law that can be used to estimate the movement of the carbonation front.

Empirically, a number of equations have been used to link carbonation rates, concrete quality, and environmental conditions. Table 3.1 summarizes some of those equations and shows the factor that has been included. In general, there is time dependence. As discussed earlier, the other factors are exposure, *w/c* ratio, strength, and CaO content (functions of cement type and its alkali content).

For example, consider this basic equation:

$$d = At^n \tag{3.2}$$

where

d is the carbonation depth in millimeters
A is the coefficient
t is the time (in years)
n is an exponent, usually equal to 0.5

A number of empirical calculations have been used to derive the values of A and n based on such variables as exposure conditions (indoor and outdoor, sheltered or unsheltered), 28-day strength, and *w/c* ratio, as shown in Table 3.1.

Schiessl (1988) showed the relation between time and the depth of corrosion under different environmental conditions (Figure 3.2). There are three curves. The first one is under laboratory conditions with a temperature of 20°C and relative humidity (RH) of about 65%, which gives a higher carbonation depth with time. The second curve puts the concrete outside the laboratory, but under a roof that provides shade; the third curve gives us less carbonation depth with time when the concrete is in a flat area outside that is not under shade.

In general, a variable (t) reflects the time in years. This factor is independent of the other factors dependent on weather conditions, concrete strength, *w/c* ratio, the

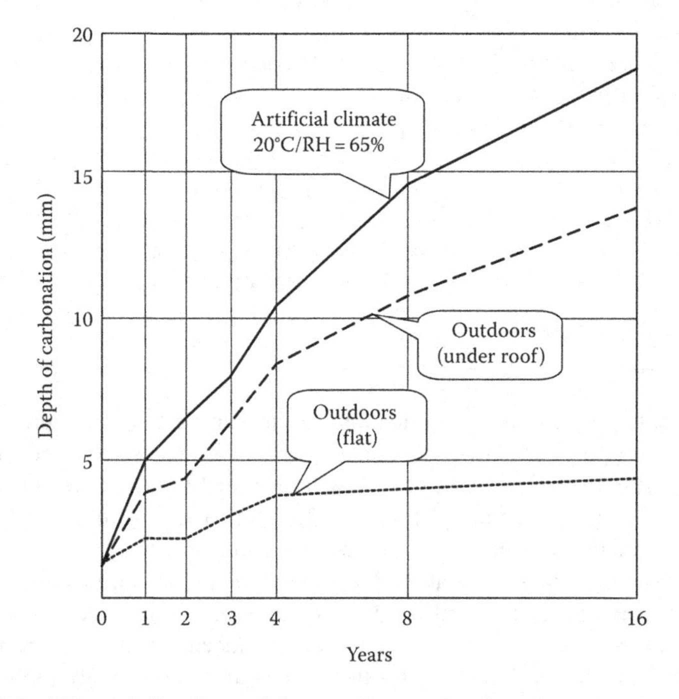

FIGURE 3.2 Effect of climatic conditions on the rate of carbonation.

content of cement used, and the resistance of concrete and calcium oxide content, which depends on the type of cement. For example, consider this equation:

$$d = A\sqrt{t} \tag{3.3}$$

The depth of carbonation is about 16 mm in 16 years in the case of weak concrete and 4 mm in 20 years for good-quality concrete. Therefore, it can be expected that the diffusion rate ranges between 0.25 and 1 mm/year.

Parrott conducted research work on the onset of steel bar corrosion; the results are made clear in Chapter 4. It is well known that, in bridges, carbonation is almost nonexistent—even after 20 years—because during the construction of bridges, more care is taken in compaction, curing, concrete mix, and concrete quality. In case of residential buildings, it will not reach 60 mm or more.

Engineers can control the carbonation process by selecting the appropriate thickness of the concrete cover and concrete mix by lowering the *w/c* ratio. Moreover, attention must be paid to quality control of construction reinforcement concrete structures. The measurement of permeability is important for determining concrete strength in order to construct durable concrete structures that can withstand carbonation.

3.3.1 Parrott's Determination of Carbonation Rates from Permeability

For a new concrete mix or structure, the prediction of carbonation rate is complicated by the lack of data to extrapolate. In a series of papers (Parrott 1994a, b; Parrott and Hong 1991), a methodology was outlined for calculating the carbonation rate from air permeability measurements using a specific apparatus. Parrott (1987) analyzed the literature and suggested that the carbonation depth D at time t is given by the following equation:

$$D = aK^{0.4}t^{n} - C^{0.5} \tag{3.4}$$

where

K is air permeability (in units of $10^{-6}\,m^2$)

C is the calcium oxide content in the hydrated cement mix for the concrete cover

$a = 64$

K can be calculated from the value at 60% relative humidity, r, by the following equation:

$$K = mK_{60} \tag{3.5}$$

where $m = 1.6 - 0.0011r - 0.0001475r^2$, or $m = 1.0$ if $r < 60$; n is 0.5 for indoor exposure but decreases under wet conditions to

$$n = 0.02536 + 0.01785r - 0.0001623r^2 \tag{3.6}$$

Therefore, increasing the concrete cover depth is required to prevent carbonation from reaching the steel. The concrete cover depth can be calculated based on the measurements of air permeability and RH.

3.4 CHLORIDE ATTACK

Chlorides can attack concrete in two ways. The first is from inside the concrete during the casting process and the second is through the movement of concrete from the outside to the inside. When casting takes place, chlorides exist in concrete due to the following:

- Seawater used in the concrete mix.
- Calcium chloride used in additives required to accelerate the setting time.
- Aggregates that contain chlorides that can be washed well.
- Additives that have a higher chloride content than that defined in the specification.
- Water used in the concrete mix that has a higher number of chloride ions than what is allowed in the specifications.

Chlorides can propagate inside concrete from the external environment because of the following:

- Concrete is exposed to seawater spray or continuous exposure to salt water.
- Salt is used to melt ice.
- The presence of chlorides in chemical substances that attack the concrete structure, such as salt storage.

In most cases, the impact of chlorides is from external sources, such as seawater spray or the use of salt in melting ice. The effect of chlorides on corrosion occurs very quickly in the case of chlorides existing in the water mixing compared to the effect of chlorides from environmental conditions surrounding the building. This often happens in an offshore structure as the concrete mix may contain seawater.

3.5 CHLORIDE MOVEMENT INSIDE CONCRETE

Similar to the carbonation rate movement, the movement of chlorides almost matches the previous rule of diffusion. But this situation is more complicated. Salt water is absorbed quickly by dry concrete. In some cases, the movement of saline water by rising capillarity is characteristic and is also due to some reactions between the concrete and chlorides, which are absorbed inside the concrete pore voids. The fundamental problem corresponding to the chloride movement is the initial concentration of chlorides when spread in the concrete. Note that the previous equation for calculating the spread of carbonation in concrete cannot be used easily in the case of chlorides. The general plot of the curve of the changing chloride concentration with depth in the concrete is explained in Figure 3.3.

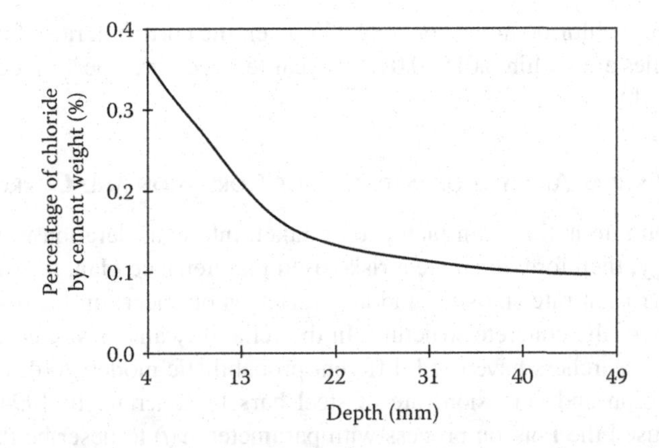

FIGURE 3.3 Chloride concentration at different depths for bridges exposed to the sea.

From this figure, we note that the concentration of chlorides is reduced rapidly whenever propagation occurs inside the concrete. It is difficult to identify chloride concentration directly on the surface, where the concentration changes with time from 0% to 100%, depending on whether the surface is dry or wet, evaporation, and many other factors. Therefore, it is common practice to measure concentration of chlorides about 5 mm from the surface.

From the previous figure, it can be seen that the chloride concentration with depth varies as a simple parabolic curve and some research is based on this (Nordic Concrete Research Group; Poulsen 1990). It is worth mentioning that the propagation of chlorides inside concrete depends not only on the diffusion factor but also on other factors influencing the starting of the propagation of the chlorides: the capillary rise phenomenon and absorption coefficient, which affect the outset at the first 5 mm from the surface.

3.6 CORROSION RATES

The carbonation transformation process, or spreading of chlorides inside concrete, is the main reason for the breaking of the passive protection of steel reinforcement and thus marking the beginning of corrosion. The deterioration of steel depends on the rate of corrosion. This, in turn, depends on many factors; in the case of carbonation, the rate of corrosion is greatly influenced by moisture-related terms. It is much lower during a decrease in the RH inside the concrete voids (less than 75%); the rate of corrosion increases significantly when the RH increases (at 95%), based on the studies by Tutti (1982). Temperature has a big impact on the rate of corrosion. For any type of corrosion, there is also a reduction (by a factor of approximately 5–10) in the corrosion rate with a 10°C reduction in temperature.

According to Schiessl (1988), the rate of corrosion in carbonated concrete is a function of RH, whether wet or dry conditions, and the chloride content. Thus, the decisive parameters controlling corrosion in carbonated concrete are associated with steady-state RH or wetting/drying cycles and conductivity increases are associated with the level of chloride in the concrete. However, the corrosion rates found in most research studies are within 0.015–0.09 mm/year (e.g., research performed in 1992 by El-Abiary et al.).

3.6.1 Statistical Analysis of Initiation of Corrosion and Corrosion Rate

The corrosion rate is the main factor to be taken into consideration in the maintenance strategy; therefore, to create a risk-based maintenance plan, it is important to know the corrosion rate statistics and any variables or uncertain factors that affect the reliability of the concrete structure. In the reliability analysis of concrete structures, many researchers have used different probabilistic models to describe initiation of corrosion and corrosion rate of steel bars in concrete. In 1994, Mori and Ellingwood used the Poisson process with parameters $\nu(t)$ to describe the initiation of corrosion following carbonation. The mean Poisson ratio is the parameter $\nu(t)$, expressed as follows:

$$\nu(w) = \begin{cases} 0 & \text{for } w < t^* \\ \nu & \text{for } w \geq t^* \end{cases} \tag{3.7}$$

where

t^* is a deterministic time, considered to be 10 years

ν is the mean initiation rate of corrosion, which is considered to be equal to 0.2/year

The typical corrosion rates of steel in various environmental conditions have been reported in recent years. According to Ting (1989), the average corrosion rate, C_r, for passive steel in concrete attacked by chlorides is about 100 µm/year. According to Mori and Ellingwood (1994), the typical corrosion rate, C_r, is a time-invariant random variable described by a lognormal distribution with a mean C_r of 50 µm/year, and a coefficient of variation V_{cr} of 50%. Because the corrosion rate changes with the environment, no accurate data are available to predict the real corrosion rate.

3.7 EFFECT OF AGE ON CONCRETE STRENGTH

Another value that affects the reliability of a concrete structure is the strength of concrete. The concrete member capacity is usually a function of concrete cross-section dimensions, the steel bar area, the concrete compressive strength, and the steel reinforcement yield strength. After some years, the concrete element will deteriorate due to corrosion of the steel bars, which will reduce the steel cross-section dimension discussed in the previous sections. The yield strength will remain the same with time, but the concrete cross-section dimension will be less effective due to concrete cracks and high reduction in the area of the section in the case of fallen concrete cover. The time required until spalling of the concrete cover will be discussed later.

The main gain with age is the increase in the concrete strength. Much research has discussed this in detail based on environmental conditions, which need to be addressed in detail in order to have an understanding about the reliability of a reinforced concrete structure. In practice, one can see corrosion of the steel bars without complete failure of deteriorated structure. With time, the concrete strength increases; this increase in strength compensates for some of the member strength due to reduction in the steel cross-section area. The concrete design approach should not depend on increase the steel reinforcement to carry most of the load as any reduction on it will be very risky, so one should not depend on gaining strength on concrete with time. However, it does need to be taken into consideration.

Much research has been done to predict concrete strength after 28 days. In the majority of cases, the tests are conducted on concrete aged 28 days, when its strength is considerably lower than what its long-term strength will be. Different methods have been suggested to predict concrete strength with age, and different codes have different recommendations for predicting this strength. For example, Baykof and Sigalof (1984) compared the gain in strength of concrete specimens stored in wet and dry conditions. They found that, in dry conditions, after 1 year there is no increase in concrete strength, as shown in Figure 3.4. On the other hand, the strength of specimens stored in a wet environment (at 15°C) is considerably increased (shown in Figure 3.4).

In Madison, Wisconsin, Washa and Wendt (1975) tested concrete specimens stored in special environmental conditions to predict the concrete strength with age and found valuable results. The specimens were moist cured for 28 days before placement outdoors on leveled ground in an uncovered, open location. Thermocouple data indicated that the outdoor compressive cylinders were subjected to about 25 cycles of freezing and thawing each winter. The RH normally varied from 65% to 100%, with an average of 75%. The annual precipitation, including snowfall, was about 32 in. Air temperatures usually ranged between 25°F and 90°F (32°C and 35°C). The average compressive strength with time is shown in Figure 3.5.

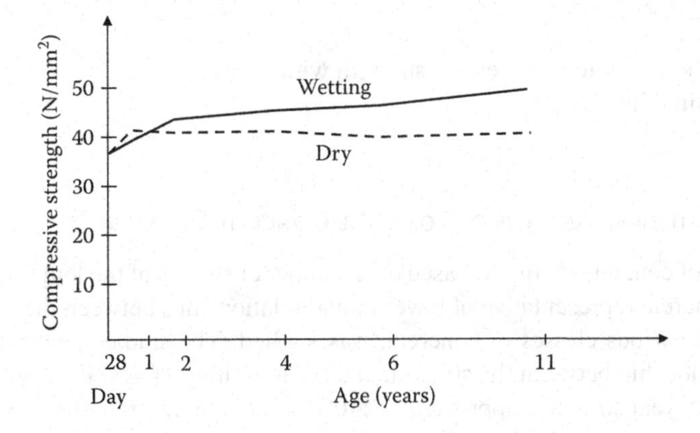

FIGURE 3.4 Variation of concrete strength with time.

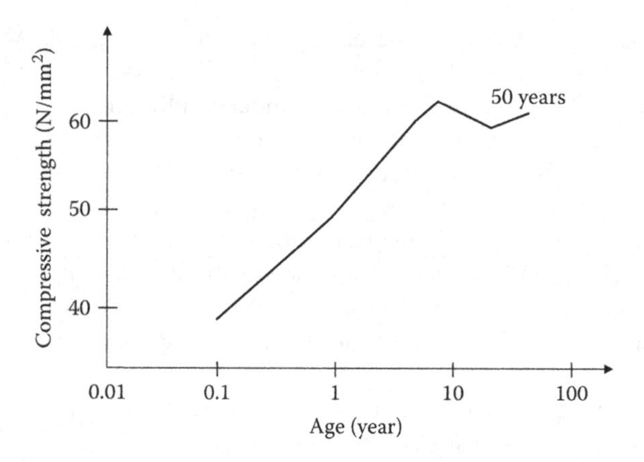

FIGURE 3.5 Variation of compressive strength with age. Adapted from Washa, G. W. and K. F. Wendt. 1975. Fifty-year properties of concrete. ACI Journal Proceedings 72(1):20–28.

Washa and Wendt (1975) concluded that the compressive strength of concrete cylinders made with cement that had a relatively low C_2S content, had a high surface area, and were stored outdoors for 50 years generally increased as a logarithm of the age for about 10 years. After 10 years, the compressive strength decreased or remained essentially the same. MacGregor (1983) used the same study and formulated an equation of the relation between compressive strength and age; in the case of the 28-day-specified compressive strength of concrete, this is equal to 281.5 kg/cm^2 (27.6 MPa). The corresponding mean of compressive strength at 28 days is 292.7 kg/cm^2 (28.7 MPa):

$$fc(t) = \begin{cases} 158.5 + 40.3ln(t)\ \text{kg/cm}^2 & t < 10\text{years} \\ 489\ \text{kg/cm}^2 & t \geq 10\text{years} \end{cases} \tag{3.8}$$

where

fc(*t*) is the concrete compressive strength with time
t is the time in days

3.7.1 Statistical Analysis of Longtime Concrete Strength

The effect of concrete maturity, based on a number of studies of the longtime strength gain of concrete representative of lower-bound relationships between the age and the strength of various classes of concrete, was studied. The studies tended to show a linear relationship between the strength and the logarithm of age. For lower-strength concrete, 25-year strengths approaching 240% of the 28-day strengths were observed by Washa and Wendt (1975). For high-strength concrete, the strength of old concrete approached 125%–150% of the 28-day strength.

3.7.2 Code Recommendations

Different codes recommend different methods to predict concrete strength for different ages. In the following sections, the gain in concrete strength with age in the Egyptian, British, and Indian codes is discussed. The compressive strength of concrete varies with age for normal concrete at moderate temperatures (Hilal 1987). The ratio of the concrete compressive strength at 28 days to that at a given concrete age may be estimated for normally and rapidly hardening portland cement according to Egyptian code of practice (ECP), as shown in the values given in Table 3.2.

According to Nevil (1983), in the past, the gain in strength beyond the age of 28 days was regarded merely as contributing to an increase in the factor of safety of the structure. Since 1957, the codes of practice for reinforced and prestressed concrete allow the gain in strength to be taken into account in the design of structures that will not be subjected to load until an older age, except when no-fines concrete (concrete with little or no fine aggregates) is used; with some lightweight aggregates, verifying tests are advisable. The values of strength given in the British code of practice CP110:1972, based on the 28-day compressive strength, are given in Table 3.3, but they do not, of course, apply when accelerators are used. This table shows that concrete continuously gains strength with time. The additional strength is about 20%–25% of the corresponding 28-day strength. Most of the additional strength is gained within the first year.

TABLE 3.2
Ratio of Fc28 to That at Age in the Egyptian Code

Age (Days)		3	7	28	90	360
Type of portland cement	Normal hardening	2.5	1.5	1.00	0.85	0.75
	Rapid hardening	1.80	1.30	1.00	0.90	0.85

Source: General Organization for Housing, Building, and Planning Research (GOHBPR 1989).

TABLE 3.3
British Code of Practice CP110:1972 Factors of Increase in Compressive Strength of Concrete with Age

Months	Age Factor for Concrete with a 28-Day Strength (MPa)		
	20–30	40–50	60
1	1.00	1.00	1.00
2	1.10	1.09	1.07
3	1.16	1.12	1.09
6	1.20	1.17	1.13
12	1.24	1.23	1.17

3.7.3 Available Statistical Parameters for Concrete Strength Considering Age

The available test results presenting the gain in concrete strength with age are discussed by Washa and Wendt (1975), who tested concrete specimens at 1, 5, 10, 25, and 50 years of age. The analysis of these data is shown in Table 3.4, where statistical parameters for the ratio of concrete strength at 1, 5, 10, 25, and 50 years and at 28 days are calculated. Baykof and Sigalof (1984) discussed the gain in concrete strength of concrete specimens stored in wet and dry conditions over time. The factor of increase in the concrete strength with age in wet and dry conditions is presented in Table 3.5. In general, buildings are always exposed to dry conditions after construction. Therefore, the case of dry conditions was taken into consideration in the study performed by El-Reedy et al. (2000), which considered the factor of increased strength after 1 year by 1.25 to the strength at 28 days with coefficient of variation equal to 0.1. This gain remained constant throughout the lifetime of the building.

TABLE 3.4
Analysis of Results of Test for Gain of Concrete Strength with Time

	28 Days	1 Year		5 Years		10 Years		50 Years	
Test No.	Fc28 N/mm²	Fc1	Fc1/Fc28	Fc5	Fc5/Fc28	Fc10	Fc10/Fc28	Fc50	Fc50/Fc28
1	28.1	36.5	1.30	52.8	1.88	55.3	1.97	53.8	1.92
2	35.4	45.9	1.30	61.7	1.75	61.2	1.73	64.8	1.83
3	30.1	40.8	1.35	52.0	1.73	55.2	1.83	53.5	1.78
4	31.8	42.3	1.33	53.0	1.67	56.4	1.77	55.7	1.75
5	43.5	57.6	1.32	63.4	1.46	73.6	1.69	73.1	1.68
6	35.1	43.9	1.25	54.9	1.57	55.4	1.58	56.5	1.61
7	36.2	51.9	1.44	55.8	1.54	59.9	1.66	62.0	1.71
8	44.1	58.4	1.33	73.3	1.66	71.3	1.62	72.4	1.64
9	36.1	51.5	1.43	62.3	1.73	63.4	1.76	55.9	1.55
10	36.2	47.2	1.31	53.7	1.48	55.6	1.54	50.2	1.39
11	49.6	61.1	1.23	70.3	1.42	66.9	1.35	65.0	1.31
12	38.5	51.8	1.34	59.1	1.53	57.5	1.49	50.6	1.31
Mean			1.33		1.62		1.67		1.62
St. dev.[1a]			0.06		0.14		0.17		0.20
COV[2b]			0.05		0.09		0.10		0.12

Source: Washa, G.W. and Wendt, K.F., *ACI J. Proc.*, 72(1), 20, 1975.

[a] Standard deviation.

[b] Coefficient of variation: equal to standard deviation divided by the mean.

TABLE 3.5
Factors of Increase in the Concrete Strength with Age in Dry and Wet Conditions

	Environment	
Age	**Dry**	**Wet**
1 Month	1.00	1.00
1 Year	1.25	1.20
2 Years	1.25	1.30
4 Years	1.25	1.40
6 Years	1.25	1.50
10 Years	1.25	1.63

Source: After Baykof, F. and Sigalof, Y. 1984. *Reinforced Concrete Structure*. Moscow, Russia: Mier.
[a] Standard deviation.
[b] Coefficient of variation: equal to standard deviation divided by the mean.

3.8 EFFECT OF CORROSION ON SPALLING OF CONCRETE COVER

As the volume of steel bar increases due to corrosion, so also as the pressure increases, cracks propagate through the cover. Experimental results have shown that, although at the early stages of crack propagation several cracks appear, by the end of the test, there is a single crack that finally breaks the cover (Nguyen et al. 2006; Ohtsu and Yosimura 1997) on the weakest side of the concrete element. When this crack appears, the internal stresses relax, which stops the propagation of other internal cracks (Nguyen et al. 2006).

Most problems that occur because of corrosion of steel in concrete are due not only to the shortage of the concrete in the steel section but also to fall of the concrete cover. Many studies and much research have been conducted to calculate the amount of corrosion occurring and causing the concrete cover to fall. It has been found that cracks may occur in cases of reduction of 0.1 mm from steel reinforcement sections and, in some cases, much less than 0.1 mm, depending on the distribution of oxides and the ability of concrete to withstand the stresses, as well as the distribution of steel.

Various efforts have been made to estimate the amount of corrosion that will cause spalling of the concrete cover. According to Broomfield (1997), the cracking is induced by less than 0.1 mm of steel bar section loss, although in some cases, far less than 0.1 mm has been needed. This is a function of the way that the oxide is distributed (i.e., how efficiently it stresses the concrete); the ability of the concrete to accommodate the stresses (by creep, plastic, or elastic deformation); and the geometry of bar distribution, which may encourage crack propagation by concentrating stresses, as in the case of a closely spaced series of bars, near the surface or at a corner where there is less confinement of the concrete to restrain cracking.

From the corrosion rate measurements, it would appear that about 10 μm section loss or 30 μm corrosion growth is sufficient to cause cracking. However, the corrosion material is a complex mixture of oxides, and hydroxides and hydrated oxides of steel have a volume ranging from twice to about six times that of the steel consumed to produce it. According to El-Abiary et al. (1992), the time, t_s, in years between initiation of corrosion and spalling of concrete is calculated from the following equation:

$$t_s = \frac{0.08 \cdot C}{d \cdot C_r} \tag{3.9}$$

where

C is the concrete cover in millimeters
d is the diameter of the steel bar
C_r is the mean corrosion rate

Few structures collapse due to corrosion in steel reinforcement because the corrosion gives some warning and deterioration is evident from the color of the concrete and the presence of the cracks. This helps in making the appropriate decision of conducting the repair process in a timely manner, as is evident in some structures in Figures 3.6–3.8.

In Figure 3.6, we clearly see the corrosion of steel stirrups: an almost completely ineffective resistance sector followed by concrete and steel corrosion—particularly, a steel bar in the corner where the corner is the fastest exhibition of the impact of carbonation and cracks on the concrete cover. Figure 3.7 shows corrosion of steel in the

FIGURE 3.6 Photo of corrosion in concrete.

FIGURE 3.7 Photo of corroded abutment bridge wall.

FIGURE 3.8 Photo of cracks in columns of elevated water tanks.

FIGURE 3.9 Photo of cracks on columns and beams.

retaining wall of a bridge. Broomfield (1997) stated that a multistory garage building fell because of corrosion. Prestressed bridges have also been found in England and Wales; the failure happened due to chlorides (also a bridge in Belgium, as stated by Woodward and Williams 1988). The spalling of concrete cover for different structure members is clearly obvious in Figures 3.7 and 3.8.

Major problems occur when there is black corrosion, particularly for structures used by the previous prestressing, as the corrosion does not form on the outer surface of steel. Often, the steel has been loaded to carry 50% or more of maximum tension resistance, and the lack of a steel reinforcement section due to corrosion strongly increases the likelihood of collapse.

Most problems, as we have said before, result from the fall of the concrete cover. For example, in New York City, a man was killed when a concrete slab bridge separated because the steel became corroded due to the use of salt to melt ice on the bridge. In Michigan, a similar accident occurred. Hence, we find that the corrosion of steel must be under control so as not to lead to falling of the concrete cover. This falling results from oxides generated on the surface of steel, which are ten times the volume of structural size, and the fact that steel buried in the concrete generates strong increases in volume. This would affect the weakest part (the concrete cover), which is where cracks form and then cause the concrete to fall.

The process of corrosion formation on steel is shown in Figure 3.8. In addition, formation of cracks and the falling of concrete cover are shown in Figure 3.9. Note that the concrete cover in the corner is more prone to falling because it has a large area for the penetration of carbon dioxide or exposure to chlorides as well as oxygen. Therefore, cracks in the concrete often form faster in this situation.

3.9 BOND STRENGTH BETWEEN CONCRETE AND CORRODED STEEL BARS

There are many research and development studies in this area, but in general, they all agree that the bond strength increases in low levels of corrosion but it will decrease rapidly in the case of high levels of corrosion as per a study conducted by Valente (2012).

There is an experimental research study conducted by Coccia et al. (2016), which revealed that a low corrosion percentage (lower than 0.5%–0.6% in mass loss) leads to an increase of the bond strength of up to about 50%–60% with respect to the soundbars; higher corrosion entities cause a sharp bond reduction; as a matter of fact for a mass loss of about 1.5%, the maximum bond stress is reduced to about 40%.

3.10 INFLUENCE OF STEEL BAR CORROSION ON THE SHEAR STRENGTH

The effect of the corroded stirrups on the shear strength was studied by Xue et al. (2012) and they conclude that if the percentage maximum local mass loss of the stirrups is below 35%, stirrup corrosion has little influence on the load-carrying mechanism, and the modified truss theory can be applied. The shear capacity can be evaluated simply by allowing for the reduction of V_s due to the sectional area loss of the stirrups. The influence of stirrup corrosion on V_c can be ignored. If the percentage maximum local mass loss of stirrup area is higher than 35%, the shear capacity decreases gradually as the corrosion level of the stirrups increases. However, there is little difference in the critical crack behavior between the corroded specimen and the sound specimen.

In the case of corroded longitudinal steel bars, Xue and Seki (2010) studied the influence of longitudinal bar corrosion on shear behavior of RC beams and they found that the bond behavior between longitudinal bars and concrete plays a very important role in the load-carrying mechanism of RC beams. As longitudinal bars corrode, the deterioration of bond strength may reduce the stiffness of RC beams and furthermore result in a transition of the load-carrying mechanism. The shear behavior of RC beams with corroded longitudinal bars is influenced not only by the corrosion level of longitudinal bars but also by the shear-span-to-effective-depth ratio. When the shear-span-to-effective-depth ratio is above 3.0, the prediction of the shear capacity of RC beams with corroded longitudinal bars using the current shear equation will result in an unsafe overvaluation.

REFERENCES

Baykof, F. and Y. Sigalof. 1984. *Reinforced Concrete Structure*. Moscow, Russia: Mier.

Broomfield, J. P. 1997. *Corrosion of Steel in Concrete*. London, UK: E & FN Spon.

Coccia, S., S. Imperatore, and Z. Rinaldi. 2016. Influence of corrosion on the bond strength of steel rebars in concrete. *Materials and Structures 49(1)*:537–551.

El-Abiary, S. et al. 1992. *Deteriorated Concrete Structures*. Cairo, Egypt: University Publishing House.

El-Reedy, M. A., M. A. Ahmed, and A. B. Khalil. 2000. Reliability analysis of reinforced concrete columns. PhD thesis. Giza, Egypt: Cairo University.

General Organization for Housing, Building, and Planning Research (GOHBPR). 1989. *Egyptian Code of Practice for Design of Reinforced Concrete Structures*. Cairo, Egypt.

Hilal, M. 1987. *Fundamentals of Reinforced and Prestressed Concrete*. Cairo, Egypt: J. Marcou & Co.

MacGregor, J. G. 1983. Load and resistance factors for concrete design. *ACI Journal 80(4)*:279–287.

Mori, Y. and B. R. Ellingwood. 1994. Maintaining reliability of concrete structures. I: Role of inspection/repair. *ASCE Journal of Structural Engineering 120(3)*:824–845.

Nevil, M. A. 1983. *Properties of Concrete*. London, UK: Pitman.

Nguyen, Q. T., A. Millard, S. Care, V. L'Hostis, and Y. Berthaud. 2006. Fracture of concrete caused by the reinforcement corrosion products. *Journal de Physique IV 136*:109–120.

Ohtsu, M. and S. Yosimura. 1997. Analysis of crack propagation and crack initiation due to corrosion of reinforcement. *Construction and Building Materials 11*(*7–8*):437–442.

Parrott, L. J. 1987. Steel corrosion in concrete: How does it occur? *Materials Protection 6*:19–23.

Parrott, L. J. 1994a. Moisture conditioning and transport properties of concrete test specimen. *Materials and Structures 27*:460–468.

Parrott, L. J. 1994b. Carbonation-induced corrosion. In Paper presented at the Institute of Concrete Technology Meeting, Reading, England, Geological Society, London, UK, November 8.

Parrott, L. J. and C. Z. Hong. 1991. Some factors influencing air permeation measurements in cover concrete. *Materials and Structures 24*:403–408.

Poulsen, E. 1990. The chloride diffusion characteristics of concrete: Approximate determination by linear regression analysis. Nordic Concrete Research, Publication No. 9, pp. 124–133.

Schiessl, P. 1988. Corrosion of steel in concrete. Report of the Technical Committee 60-CSC, London, UK: Chapman and Hall.

Ting, S. C. 1989. *The Effects of Corrosion on the Reliability of Concrete Bridge Girder. PhD thesis*. Ann Arbor, MI: University of Michigan.

Tutti, K. 1982. *Corrosion of Steel in Concrete*. Stockholm, Sweden: CBI Swedish Cement and Concrete Institute.

Valente, M. 2012. Bond strength between corroded steel rebar and concrete. *IACSIT International Journal of Engineering and Technology 4*(*5*):653–656.

Washa, G. W. and K. F. Wendt. 1975. Fifty-year properties of concrete. *ACI Journal Proceedings 72*(*1*):20–28.

Woodward, R. J. and F. W. Williams. 1988. Collapse of Ynes-y-Gwas Bridge, West Glamorgan. *Proceedings of the Institution of Civil Engineers Part I 84*:635–669.

Xue, X., S. Hiroshi, and Y. Song. 2012. Influence of stirrup corrosion on shear strength of RC beams. *Applied Mechanics and Materials 204–208*:3287–3293.

Xue, X. and H. Seki. 2010. Influence of longitudinal bar corrosion on shear behavior of RC beams. *Japan Concrete Institute Journal of Advanced Concrete Technology 8*(*2*):156.

4 Assessment Methods for Reinforced Concrete Structures

4.1 INTRODUCTION

In Chapter 2, the corrosion of steel buried in concrete was studied, and it was found that the corrosion shows the same characteristics in terms of its impact on steel in any situation. The causes of corrosion or its types that affect the steel have been studied in detail in Chapter 3. When a concrete structure shows signs of corrosion, it should be repaired, but before that, the building must be evaluated with high accuracy, and the environmental conditions surrounding it must be identified.

After performing risk assessment on the structure, we can then identify the parts of the concrete elements that will be repaired and determine the method of repair. The choice of a correct and technically accurate assessment method to determine the degree of building risk will lead to an effective repair process. The process of building assessment is the preliminary step; it diagnoses the defects in a structure as a result of corrosion and identifies the causes that have led to the corrosion. Therefore, we will have to resolve two issues: one to determine the method of repair and the other to define a reasonable method to protect steel bars from corrosion in the future.

The process of assessment of structures often takes place in two stages. The first stage involves the initial assessment of the building, definition of the problem, and the development of a plan for the detailed evaluation of the structure. A detailed assessment of the building will carefully define the problems and their causes; at this stage, a detailed inspection of the whole building will be carried out. A basis for the inspection of a building has been established by evaluating it through technical report 26 (Concrete Society 1984) as well as through American Concrete Institute (ACI) standards.

There are many reasons for the deterioration of reinforced concrete structures, ranging from the presence of cracks to the falling of the concrete cover. Among these reasons are increasing stresses, plastic shrinkage, frost or cracked concrete in the case of plastic concrete, moving of the wood form during construction, or the presence of aggregates, which have the capability to shrink or deteriorate as a result of the interaction of the aggregates with alkalinity. Some structures, such as reinforced concrete pipelines in sanitary projects and underground concrete structures like tunnels, have surfaces that are in direct contact with the soil. In these cases, the carbonate present in the water attacks the concrete. Although this type of deterioration is

DOI: 10.1201/9781003407058-4

important to consider, we will focus here on corrosion as a result of the exposure of concrete structures to the atmosphere.

As we have stated previously, there are several types of cracks. Therefore, the engineer responsible for performing the evaluation must be highly experienced in that area so that he can specify precisely why the deterioration in concrete occurs. A wrong diagnosis will result in the wrong repair, which will cause heavy financial losses and have an impact on the safety of the structure.

The evaluation of a structure is not only a structural assessment but also an assessment of the state of the concrete in terms of the presence of corrosion and the rate of corrosion and collapse of concrete. From previous information, we have the ability to decide whether the concrete member can withstand loads. In addition, a case study will be done to assess the member to ascertain whether a deflection of more than the allowable limit will cause cracks, reduction in cross-sectional area, or falling concrete cover.

4.2 PRELIMINARY INSPECTION

During this inspection, it is necessary to clearly identify the geographical location of the building, the nature and circumstances of the weather conditions surrounding the building, the method of construction of the building, its structural system, and the method of loading. This assessment is often performed by a preliminary visual inspection concentrated on cracks and the fall of concrete. It also focuses on collecting data regarding the thickness of the concrete cover and the nature of the structure in terms of the quality of concrete and of construction as well as the type of the structural system—whether it consists of a beam and column system, or slab-on-load-bearing masonry, or prestressed or precast concrete. It is necessary to perform some simple measurements—for example, determining the extent of the carbonation transformation in the concrete. A sample can be taken from concrete that has fallen and laboratory tests performed.

The safety of the structure must be calculated precisely, especially after reducing the cross-sectional area of steel due to corrosion; the fall of the concrete cover also reduces the total area of the concrete member. The aforementioned two areas are the main coefficients that directly affect the capacity of the concrete member to carry load.

4.3 DETAILED INSPECTION

The purpose of the detailed inspection is to determine accurately whenever possible the degree of seriousness and the deterioration of the concrete. Therefore, we need to know the amount of the collapse that has occurred, the cause of the deterioration in the concrete, and the amount of repair that will be needed. These must be defined precisely at this stage, as such quantities have to be put forward to contractors for repair. At this stage, we will need to have a detailed knowledge of the reasons for the collapse and the contractor should possess the capability for performing the failure analysis technique.

Initially, visual inspection may be carried out in conjunction with the use of a small hammer before carrying out other measurements to determine the depth of carbon transformation in the concrete, as well as the degree of steel corrosion in concrete and how much it extends into the steel bars. Moreover, in this stage, it is important to define the degree of concrete electric resistivity to predict the corrosion rate. All these required measurements are discussed in this chapter. As should be known, the weather conditions affecting the building are the main factors that affect the measurement readings; they also affect the selection of the method of repair.

In general, as per ACI 562-19 (2019), which presents that in the detailed inspection, the condition of the structure in general should cover these seven points:

1. The physical condition of the structural members to examine the extent and location of deterioration or distress.
2. The adequacy of continuous load paths through the primary and secondary structural members to provide for life safety and structural integrity.
3. As-built information required to determine appropriate strength reduction factors.
4. Structural members' orientation, displacements, construction deviations, and physical dimensions.
5. Properties of materials and components from available drawings, specifications, and other documents; or by testing of existing materials.
6. Additional considerations, such as proximity to adjacent buildings, load-bearing partition walls, and other limitations for rehabilitation.
7. Information needed to assess lateral-force-resisting systems, span lengths, support conditions, building use and type, and architectural features.

4.4 METHODS OF STRUCTURE ASSESSMENT

Several methods assess the structure in terms of the extent of corrosion in steel reinforcement bars and its impact on the whole structure. The first and most important method is visual inspection because it is not expensive and is easily performed. This is followed by other methods that require certain skills and are often used in the case of structures of special importance that need expensive repair. Therefore, the use of sophisticated, highly accurate technology is required to identify the degree of corrosion of the steel reinforcement in concrete because this affects the total cost of the structure—for example, concrete bridges or special constructions, such as parking garages or tunnels.

Each of the measurement tools used has a specific accuracy with advantages and disadvantages. In general, the process of assessing the structure for corrosion attack must be conducted by a person who has an acceptable amount of experience in evaluating the process of corrosion because experience is the master key and plays a big role in successful assessment. However, some devices need specialized experienced knowledge of their use as well as of the accuracy and environmental factors that affect the equipment readings and how to overcome them.

In Table 4.1, every method is identified, along with the user's ability to work with that method and the performance rate. This will help to estimate the cost of the inspection for evaluating the building and the performance rates will assist us in performing the inspection on schedule. These measurements will determine the cause

TABLE 4.1
Practical Methods to Evaluate Concrete Structures

Methods	Inspection	User	Approx. Performance Rate
Visual inspection	Surface defects	General	1 m^2/second
Chain or hammers	Void behind cover	General	0.1 m^2/second
Concrete cover measurement	Distance between steel bars and concrete surface	General	One reading every 5 minutes
Phenolphthalein	Carbonation depth	General	One reading every 5 minutes
Half cell	Evaluate corrosion risk	Expert	One reading/5 seconds
Linear polarization	Corrosion rate	Expert	One reading in 10–30 minutes
Radar	Defects and steel location	Expert	1 m/second by using car or 1 m^2 in 20 seconds[a]

of corrosion and the degree of passive protection layer for the steel bars as well as the expected corrosion rate in every part of the building. From this information, we can determine the kind of repair the structure needs, the type of construction, the calculation of the quantity of materials, and a survey of the repair required to strengthen the concrete member.

4.4.1 Visual Inspection

Visual inspection is the first step in any process of technical diagnosis and generally is performed by first viewing the structure as a whole and then concentrating on the general defects. It is necessary to define precisely the deterioration due to corrosion and the extent of corrosion of the steel reinforcement. As we have stated before, the assessment of the building must be performed by an expert because the cracks in the concrete structure may not be the cause for corrosion—corrosion is not the only factor that causes cracks. However, it is the main reason for a major deterioration of structures. Later, Figure 4.2a–c illustrates various forms and causes of cracks in beams, slabs, and columns.

In general, there are many causes for cracks in concrete. They may affect only the appearance or may indicate significant structural distress or a lack of stability. Cracks may either represent the total extent of the damage or may point to problems of greater magnitude. Their significance depends on the type of structure, as well as the nature of the crack. For example, cracks that are acceptable for building structures may not be acceptable in water-retaining wall structures.

The proper repair of cracks requires a knowledge of the causes of the cracks and selection of repair procedures that take them into account; otherwise, the repair may only be temporary. Successful long-term repair procedures must prevent the causes of the cracks in addition to eliminating the cracks themselves. Cracks may occur in plastic concrete or in hardened concrete. Cracks in plastic concrete occur as plastic shrinkage cracking, settlement cracking, and, after hardening, dry shrinkage cracking. The following sections will illustrate these types of cracking based on the ACI code.

4.4.1.1 Plastic Shrinkage

Cracking caused by plastic shrinkage in concrete occurs most commonly on the exposed surfaces of freshly placed floors and slabs or other elements with large surface areas when they are subjected to a very rapid loss of moisture caused by low humidity and wind or high temperature or both. Plastic shrinkage usually occurs prior to final finishing, before curing starts. When moisture evaporates from the surface of freshly placed concrete faster than it is replaced by curing water, the surface of the concrete shrinks. Due to the restraint exerted by the concrete on the drying surface layer, tensile stresses develop in the weak, stiffening plastic concrete, resulting in shallow cracks that are usually not short and run in all directions. In most cases, these cracks are wide at the surface. They range from a few millimeters to many meters in length and are spaced from a few centimeters to as much as 3 m apart. Plastic shrinkage cracks may extend the full depth of elevated structural slabs.

Since cracking due to plastic shrinkage is due to a differential volume change in the plastic concrete, successful control measures require a reduction in the relative volume change between the surface and other portions of the concrete. There are many methods and techniques to prevent this type of crack in the case of a rapid loss of moisture due to hot weather and dry winds. These methods include the use of fog nozzles to saturate the air above the surface and using plastic sheeting to cover the surface between the final finishing operations. In many cases, during construction, it is preferable to use wind breakers to reduce the wind velocity; sunshades to reduce the surface temperature are also helpful. In addition, it is good practice to schedule flatwork after the walls have been erected.

4.4.1.2 Settlement Crack

After initial placement, vibration, and finishing, concrete has a tendency to continue to consolidate. During this period, plastic concrete may be locally restrained by reinforcing steel, a prior concrete placement, or formwork. This local restraint may result in voids and/or cracks adjacent to the restraining element, as shown in Figure 4.1. When associated with reinforcing steel, settlement cracking increases with increasing bar size, increasing slump, and decreasing cover. The degree of settlement cracking will be magnified by insufficient vibration or the use of leaking or highly flexible forms. Proper form design and adequate vibration, provision of sufficient time intervals between the placement of concrete in slabs and beams, the use of the lowest possible slump, and an increase in concrete cover will reduce settlement cracking.

4.4.1.3 Shrinkage from Drying

The common cause of cracking in concrete is shrinkage due to drying. This type of shrinkage is caused by the loss of moisture from the cement paste constituent, which can shrink by as much as 1% per unit length. Unfortunately, aggregation provides an internal restraint that reduces the magnitude of this volume change to about 0.05%. Upon wetting, concrete tends to expand.

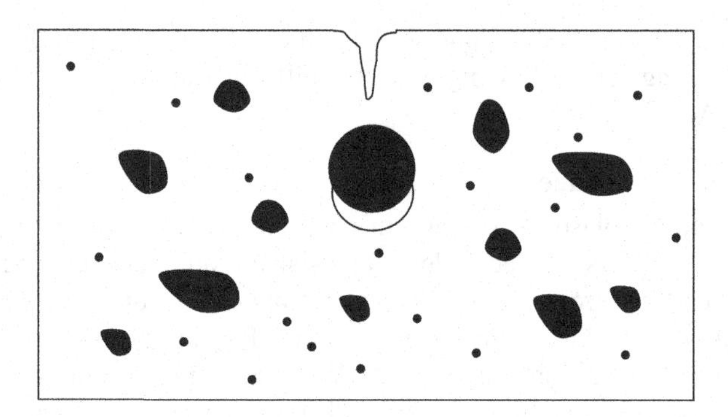

FIGURE 4.1 Settlement cracking. The effects of settlement cracking in concrete on corrosion initiation and rate of reinforcing steel in the presence of chlorides were studied by Darwin (2022) by using uncracked specimens, cracked beam (CB) specimens with 0.012 in. (0.3 mm) artificial cracks directly above reinforcing steel, and settlement cracking (SC) specimens in which cracks with widths ranging from 0.001 to 0.004 in. (0.025–0.10 mm) form in plastic concrete over reinforcing bars. The earliest corrosion initiation was observed in the CB specimens, followed, in turn, by the settlement cack and then uncracked specimens. It is found that, although narrow, settlement cracks can lead to early initiation of corrosion—on average of less than half the time than for uncracked concrete. Relative to uncracked concrete, specimens with settlement cracks exhibited a 30% increase in corrosion rate while specimens with the artificial 0.012 in. (0.3 mm) crack exhibited an over 200% increase in corrosion rate.

These moisture-induced volume changes are characteristic of concrete. If the shrinkage of concrete could take place without any restraint, then the concrete would not crack. It is the combination of shrinkage and restraint, which is usually provided by another part of the structure or by the subgrade, which causes tensile stresses to develop. When the tensile stresses of concrete are exceeded, it will crack. Cracks may propagate at much lower stresses than are required to cause crack initiation.

In massive concrete structure elements, tensile stresses are caused by differential shrinkage between the surface and the interior concrete. The larger shrinkage at the surface causes cracks to develop that may, with time, penetrate more deeply into the concrete. The magnitude of the tensile stresses is influenced by a combination of the following factors:

- Amount of shrinkage.
- Degree of restraint.
- Modulus of elasticity.
- Amount of creep.

The amount of drying shrinkage is influenced mainly by the amount and type of aggregate and the water content of the mix. The greater the amount of aggregate, the smaller is the amount of shrinkage. The higher the stiffness of the aggregate, the more effective it is in reducing the shrinkage of the concrete. This means that concrete containing a sandstone aggregate has a higher shrinkage rate—about twice that of concrete containing granite, basalt, or limestone. The higher the water content, the greater is the amount of shrinkage from drying.

Surface crazing on walls and slabs is an excellent example of shrinkage due to drying on a small scale. Crazing usually occurs when the surface layer of the concrete has a higher water content than that of the interior concrete. The result is a series of shallow, closely spaced fine cracks. Shrinkage due to drying can be reduced by using the maximum amount of aggregate practically possible in the mix. The lowest water-to-cement ratio is important to avoid this type of shrinkage. A procedure that will help reduce settlement cracking, as well as drying shrinkage in walls, consists in reducing the water content in concrete as the wall is placed from the bottom to the top. Using this procedure, bleed water from the lower portions of the wall will tend to equalize the water content within the wall. To be effective, this procedure needs careful control and proper consolidation.

Cracking due to shrinkage can be controlled by using appropriately spaced contraction joints and accurate steel detailing. It may also be controlled by using shrinkage-compensating cement.

4.4.1.4 Thermal Stresses

The temperature differences within a concrete structure may be due to cement hydration or changes in ambient temperature conditions or both. These temperature differences result in differential volume changes. The concrete will crack when the tensile strains due to the differential volume changes exceed their tensile strain capacity.

The effects of temperature differentials due to the hydration of cement are normally associated with massive concrete formations such as large columns, piers, beams, footing, retaining walls, and dams, while temperature differentials due to changes in the ambient temperature can affect any structure. Considering thermal cracking in massive concrete forms, production procedures for portland cement cause it to heat as it hydrates, causing the internal temperature of concrete to rise during the initial curing period. The concrete rapidly gains both strength and stiffness as cooling begins. Any restraint of the free contraction during cooling will result in tensile stress. Tensile stresses developed during the cooling stage are proportional to the temperature change, the coefficient of thermal expansion, the effective modulus of elasticity, and the degree of restraint. The more massive the structure, the greater is the potential for temperature differential and degree of restraint.

Procedures to help reduce thermally induced cracking include reducing the maximum internal temperature, delaying the onset of cooling, controlling the rate at which the concrete cools, and increasing the tensile strain capacity of the concrete.

Hardened concrete has a coefficient of thermal expansion that may range from 7 to 11×10^{-6}/°C, with an average of 10×10^{-6}/°C. When one portion of a structure is subjected to a temperature-induced volume change, the potential for thermally induced cracking exists. Designers must give special consideration to structures in which some portions are exposed to temperature changes while other portions of the structure are either partially or completely protected. A drop in temperature may result in cracking of the exposed element, while increase in temperature may cause cracking of the protected portion of the structure. Therefore, the designer must allow the movement of the structure by recommending the use of contraction joints and providing the correct detailing to it.

Structures that experience high differences in temperature are usually concrete structures built in areas near a desert, where temperatures can vary greatly between afternoon and midnight. Moreover, countries with high temperatures usually have air-conditioning inside buildings, so there will be a high probability for formation of cracks due to the difference in temperatures inside and outside the building. The designer should take these stresses into consideration.

4.4.1.5 Chemical Reaction

As shown in Figure 4.2a, starlike cracks in the concrete surface are an indication of a chemical reaction. This reaction occurs when an aggregate is present that contains

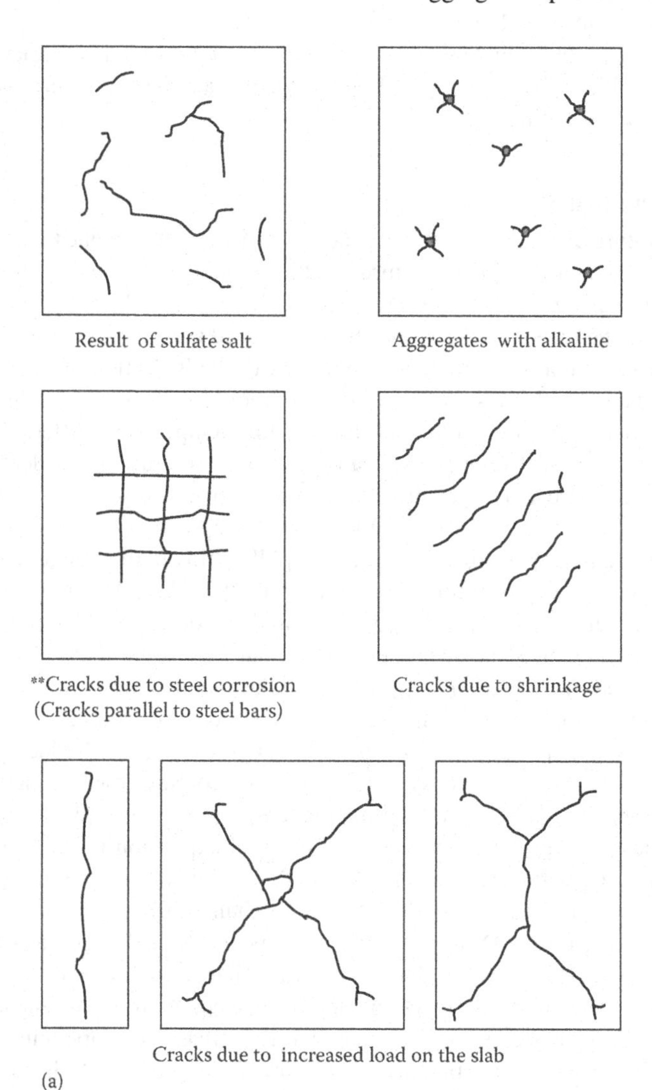

FIGURE 4.2 Types of cracks in reinforced concrete: (a) slabs, (b) beams, (c) columns.

(Continued)

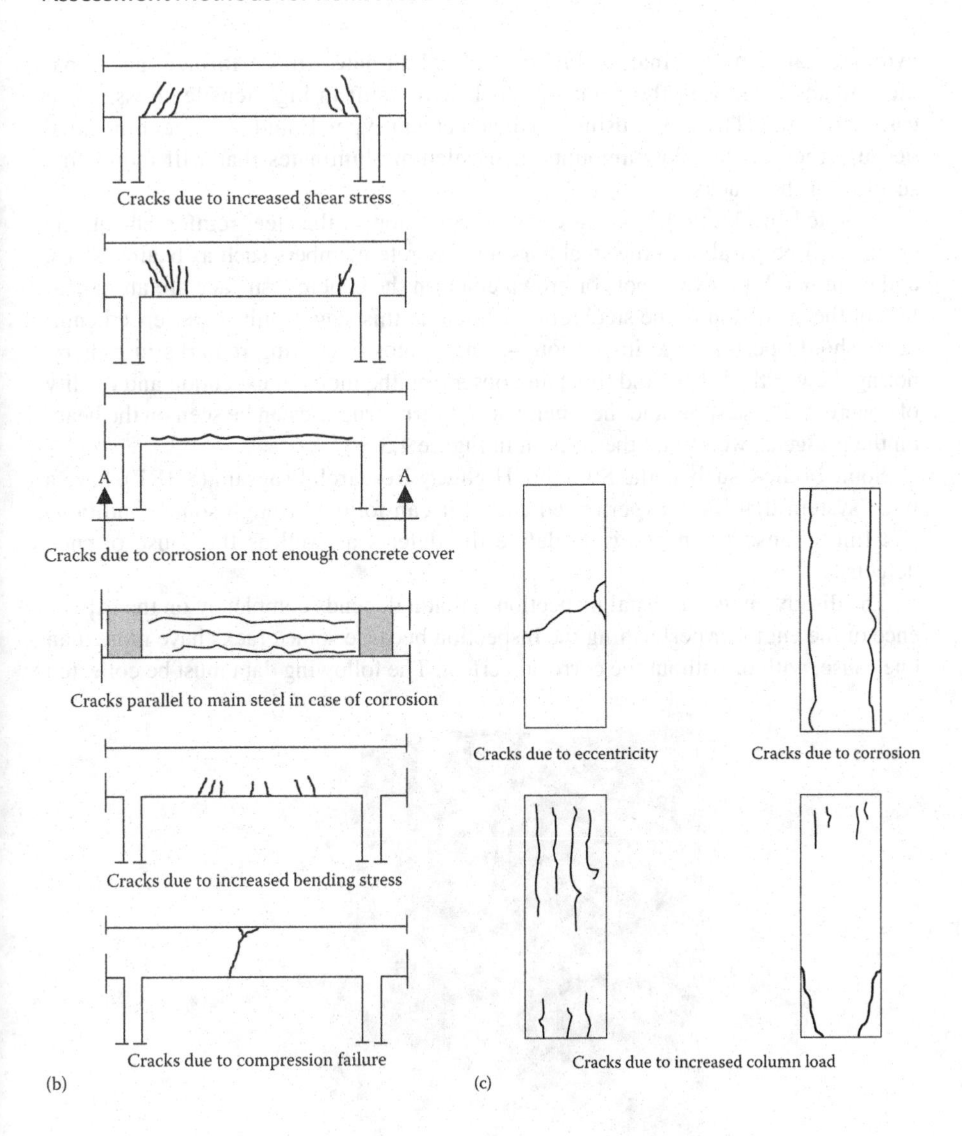

FIGURE 4.2 ***(Continued)*** Types of cracks in reinforced concrete: (a) slabs, (b) beams, (c) columns.

active silica and alkalis derived from cement hydration, admixtures, or external sources, such as curing water, groundwater, or alkaline solutions stored or used in the finished structure.

The alkali–silica reaction results in the formation of a swelling gel, which tends to draw water from other portions of the concrete. This causes local expansion and accompanying tensile stresses and may eventually result in the complete deterioration of the structure.

Groundwater that has sulfate poses a special durability problem for concrete. The sulfate penetrates hydrated cement paste so that it comes in contact with hydrated calcium aluminates. Calcium sulfoaluminate will be formed, accompanied by an increase in the volume, which will result in high tensile stresses that cause cracking. Therefore, using portland cement types II and V is recommended because they contain low amounts of tricalcium aluminates that will reduce the severity of the cracks.

As noted in Figure 4.2a–c, in cases of corrosion in the steel reinforcement, the cracks will be parallel to the steel bars for concrete members such as beams, slabs, and columns. Moreover, spots of brown color on the concrete surface are an indication of the corrosion in the steel reinforcement at this stage. Only experienced engineers should perform the inspection—using photos; recording remarks in a diary, noting the weather, date, and time; and observing the mode of execution and quality of concrete. These spots and the emergence of infrastructure can be seen on the beam on the bridge as well as on the column in Figure 4.3.

Some bodies, such as the Strategic Highway Research Program (SHRP), have a fixed system that the inexperienced engineer can follow through some conditions and limitations; it can precisely define the defects as well as the cause of such defects.

The disadvantage of visual inspection is that it depends completely on the experience of the engineer performing the inspection because some cracks have more than one cause, with or without the corrosion effect. The following data must be collected

FIGURE 4.3 Brown spots on bridge column and girder.

before performing the visual inspection and must be stated in the structure assessment report:

- Construction year.
- Contractor and engineering office data.
- Structural system.
- Drawings and project specifications.
- Construction method.
- Concrete tests, if available.
- Environmental conditions (near the sea or a chemical factory).
- Historical data and events available.

The owner must submit these data to the engineer, if he is a third-party expert. In most cases, all these data cannot be collected because, logically speaking, a building may be over 20 or 30 years old; therefore, the drawings and specifications will not be available because computers and software drawing programs were not common 20 years ago. All the hard-copy drawings, in most cases, are not available or are in bad condition. If the owner has a big organization with an engineering department, the engineers may have a document control system and the consulting engineer can find the drawing. This can be possible for office and industrial buildings, but for a residential building, the engineer will be very lucky if the owner has these drawings.

The construction year is a very critical piece of information because the codes and specifications that were applied in this year can be obtained from this, although this depends on the engineer's experience. Knowing the construction year, the contractor who constructed the building, and the engineering office that provided the engineering design, the engineer can imagine the condition of the building based on the reputation of the contractor and the engineering office.

The construction system can be ascertained by visual inspection, the drawing, or talking with the user of the building and the engineering firm that shared reviewing the engineering design documents or supervising the construction.

Information on the environmental conditions will be provided by the owner. The engineer can see it on-site, collect the data about the environmental conditions from other sources, and take into consideration all the data, including temperatures in summer and winter, morning and night, and the wind and wave effects.

The owner must be sure that all the data are delivered to the consultant who does the inspection because all data are valuable to him. He needs to remember the motto "garbage in, garbage out" and verify all the data.

In industrial facilities, the plant conditions and mode of operations are not clear to the consultant engineer, so this information must be delivered to him. Machine vibrations, heat of liquid in pipelines, and so on have more impact on the concrete structure assessment results. When data are collected from a processing plant or factory, it is necessary to consider that the mode of operations may change from time to time. Therefore, all these histories must be delivered to the engineer conducting the evaluation, who is usually a structural engineer unfamiliar with and unable to imagine these situations.

If the data for winds and waves do not exist or are not known, they must be obtained from a third party specializing in meteorological or oceanological information. These data are very important in cases of offshore structures or marine structures. Their assessment depends on these data if winds and waves represent the greatest loads affecting the structure.

The concrete compressive strength test is usually performed in any building by cylindrical specimens, as in the American code, or cubic specimens, as in the British standard. However, the problem is to find the data after 20 years or more, depending on the age of the building. If they are not available, the test must be performed using an ultrasonic pulse velocity test, a Schmidt hammer, or a core test.

4.4.2 Concrete Test Data

The following sections describe tests that should be performed to obtain data on concrete strength if these data are not available. There are variations among these tests concerning accuracy, effects on the building, and cost. Therefore, it is the responsibility of the consulting engineer to decide whether any of these tests is suitable for the building to be evaluated. Moreover, there are some precautions to be taken for these tests: When the samples are taken, they must be based on the visual inspection results that were done before. In addition, when samples are taken or the loading test is conducted, it must be accompanied by a suitable temporary support to the structure and the adjacent members. These temporary support locations and strengths should be suitable to withstand the load in the case of weakness of the structure member that is being tested. These tests will follow ACI 228-89-R1 (1989) and BS 1881 (1971a,b, 1983).

4.4.2.1 Core Test

This test is considered to be one of the semi-destructive tests. It is very important and popular to study the safety of a structure as a result of change in the system of loading or deterioration of structure as a result of accidents, such as fire or weather factors. It is also used when temporary support is needed for repair and no accurate data about concrete strength are available. This test is not too expensive and is the most accurate test to determine the strength of concrete actually carried out.

The core test (American Concrete Society 1987; ASTM C 42-90m 1990; BS 1881 1983) is done by cutting cylinders from the concrete member, which could affect the integrity of the structure. Therefore, the required samples must be taken according to the standard, as the required number will provide adequate accuracy of the results without weakening the building.

In our case, for a structure that has deteriorated due to corrosion of steel bars, the structure has lost most of its strength due to the reduction of the steel cross-sectional area. Thus, more caution must be used when performing this test and selecting the proper concrete member on which to perform it so that there will be no effect on the building from a structural point of view. Therefore, the codes and specifications provide some guidance to the number of cores to test; these values are as follows:

Volume of concrete member $(V) \leq 150\,m^3$ takes 3 cores.
Volume of concrete member $(V) > 150\,m^3$ takes $3 + (V - 150/50)$ cores.

TABLE 4.2
Number of Cores and Deviation in Strength

Number of Cores	Deviation Limit between Expected Strength and Actual Strength (Confidence Level 95%) (%)
1	+12
2	+6
3	+4
4	+3

[a] Add more time than that during schedule plan preparation.

The degree of confidence of the core test depends on the number of tests, which must be the minimum possible. The relation between the number of cores and confidence is found in Table 4.2.

Before the consulting engineer chooses the location of the sample, he must define the location of the steel bars to assist in selecting the location of the sample away from them, if possible, in order to avoid taking samples containing steel reinforcement bars. He must carefully determine the places to preserve the integrity of the structure; therefore, this test should be performed by an experienced engineer who conducts such an experiment taking precautions, determining the responsibility of individuals, and accurately reviewing the nondestructive testing that has been conducted. Figures 4.4 and 4.5 present the process of taking the core from a reinforced concrete bridge girder.

4.4.2.1.1 Core Size

Note that the permitted diameter is 100mm in the case of a maximum aggregate size of 25 and 150mm in the case of a maximum aggregate size not exceeding 40mm. It is preferable to use 150mm diameter whenever possible, as it gives more accurate results. This is shown in Table 4.3, which represents the relationship between the

FIGURE 4.4 Taking a core sample.

FIGURE 4.5 How to take a core sample.

TABLE 4.3
Core Size with Possible Problem

Test	Diameter (mm)	Length (mm)	Possible Problem
1	150	150	May contain steel reinforcement
2	150	300	May cause more cutting depth to concrete member
3	100	100	Not allowed if maximum aggregate size is 25 mm May cut with depth less than required
4	100	200	Less accurate data

dimensions of the sample and potential problems. This table should be considered when choosing a reasonable core size. Some researchers have stated that the core test can be done with a core diameter of 50 mm in the case of a maximum aggregate size and not more than 20 mm overall; note that small core sizes give results different from those of large sizes.

Because of the seriousness of the test and inability to take high numbers of samples, the gathering of the sample should be well supervised. Moreover, the laboratory test must be certified and the test equipment must be calibrated and a certificate of calibration from a certified company must be held.

Sample extraction uses pieces of a cylinder, which differ depending on the country. Cylinders are equipped with a special alloy mixture with diamond powder to feature pieces in the concrete during the rotation of the cylinder through the body. As a precaution, sampling methods should match and bring appropriately consistent pressure to be borne; this depends on the expertise of the technician.

After that, the core will be filled with dry concrete of suitable strength or grouting will be poured; the latter is a popular method. Another solution depends on using epoxy and injecting it into the hole and then inserting a concrete core of the same size to close the hole. No matter which method is chosen, the filling must be done soon after the cutting. The filling material will be kept by the technician who does the cutting, as this core affects the integrity of the structure.

The lab must examine and photograph every core and note gaps identified within the core as small voids, if they measure between 0.5 and 3 mm; voids, if they measure between 3 and 6 mm; or large voids, if they measure more than 6 mm. The lab also examines whether the core is nesting and determines the shape, kind, and color gradient of aggregates, as well as any apparent qualities of the sand. In the laboratory, dimensions, weight of each core, density, steel bar diameter, and distance between the bars will be measured.

4.4.2.1.2 Sample Preparation for Test

After the core is cut from the concrete element, the sample is processed for testing by leveling the surface of the core. A core that has a length of not less than 95% of the diameter and not more than double the diameter is taken. Figure 4.6 shows the shape of the core sampling after cutting from the concrete structure member directly.

For leveling the surface, a chain saw, spare concrete, or steel cutting disk is used. After that, the two ends of the sample are prepared by covering them with mortar or sulfide and submerging the sample in water at a temperature of 20°C ± 2°C for at least 48 hours before testing it. The sample is put through a machine test and an influence load is applied gradually at the rate of a regular and continuous range of 0.2–0.4 N/mm^2, until it reaches the maximum load at which the sample has been crushed.

FIGURE 4.6 Shape of core sample.

The estimated actual strength for a cube can be determined by knowing the crushing stress obtained from the test and using the following equation, as λ is the core length divided by its diameter. In the case of a horizontal core, the strength calculation will be

$$\text{Estimated actual strength for cube} = 2.5/(1/\lambda) + 1.5 \times \text{Core strength}$$

where
λ = core length/core diameter.

In the case of a vertical core, the strength calculation is as follows:

$$\text{Estimated actual strength for cube} = 2.3/(1/\lambda) + 1.5 \times \text{Core strength} \tag{4.1}$$

ACI 562-19 presents the following equation to provide the equivalent strength as per cores data, which are as follows:

$$f_{ceq} = 0.9\overline{f}_c\left[1 - 1.28\sqrt{\frac{(k_c V)^2}{n} + 0.0015}\right]$$

where $\overline{f}_c$ is the average core strength, as modified to account for the diameter, length-to-diameter ratio, and moisture condition of the core (following ASTM C42 procedures); V is the coefficient of variation of the core strengths (a dimensionless quantity equal to the sample standard deviation divided by the mean); n is the number of cores taken; and kc is the coefficient of variation modification factor, as obtained from Table 4.4.

TABLE 4.4
Concrete Coefficient of variation Modification Factor Kc

n	Kc
2	2.4
3	1.47
4	1.28
5	1.20
6	1.15
8	1.10
10	1.08
12	1.06
16	1.05
20	1.03
25 or more	1.02

In the case of existing steel in the core perpendicular to the core axis, the previous equations will be multiplied by the following correction factor:

$$\text{Correction factor} = 1 + 1.5(s\phi)/(LD) \quad (4.2)$$

where
L is the core length
D is the core diameter
s is the distance from steel bar to edge of core
ϕ is the steel bar diameter

Cores are preferred to be free of steel; if steel is found, it is necessary to use the correction factor, taking into account that it is taken only in the event that the value ranges from 10% to 5%. We must agree to use the results of the core, but if the correction factor is more than 10%, the results for the core cannot be trusted and another core should be taken.

When examining the test results, certain points must be taken into account. Before testing, the sample should be submerged in water; this leads to a decrease in strength of up to about 15% for the strength of dry concrete.

According to BS6089, there is a difference between estimated in situ cube strength and design strength. The level of in situ cube strength that may be considered acceptable in any particular case is a matter for engineering judgment but should not normally be less than 1.2 times the design strength, whereas the design strength is f_{cu}/γ_m and $\gamma_m = 1.5$. So $f_{cu} = 1.25\times$ estimated in situ concrete strength. For example, if as per the project specification the concrete strength after 28 days $f_{cu} = 40$ MPa, the estimated in situ concrete strength as per Equation 4.1 shall not be less than 32 MPa.

In the case of prestressed concrete, the concrete strength is acceptable if the average strength of the cores is at least 80% of the required strength and the calculated strength for any core is less than 75% of the required strength.

4.4.2.2 Rebound Hammer

This is nondestructive testing, so it is useful in determining the estimated concrete compressive strength. This is the most common test, as it is easy to perform and is very inexpensive compared with other tests; however, it gives less-precise outcomes of data results. This test relies on measuring the concrete strength by measuring the hardening from the surface. It will be able to identify the concrete compressive strength of the concrete member by using calibration curves of the relationship between reading the concrete hardening and concrete compressive strength. Figures 4.7–4.9 present different types of rebound (Schmidt) hammer from different ways of reading the results.

Rebound hammers most commonly give an impact energy of 2.2 N/mm^2. There is more than one way to show results, based on the manufacturer; in some cases, the reading will be an analog or digital number or connected to a memory device to record the readings. Before it is used, the rebound hammer should be inspected using

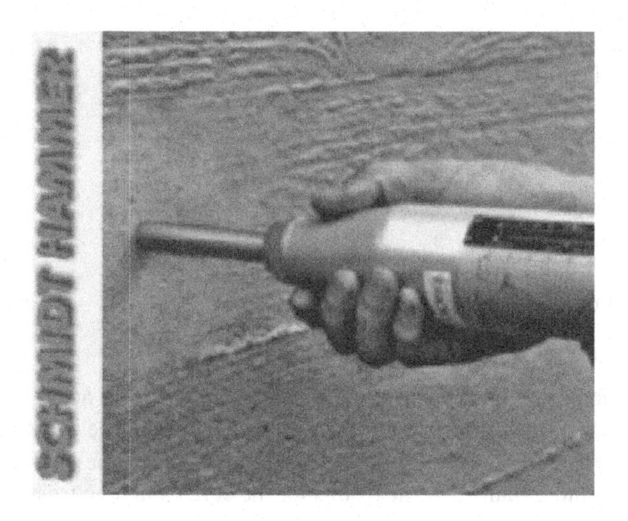

FIGURE 4.7 Rebound hammer.

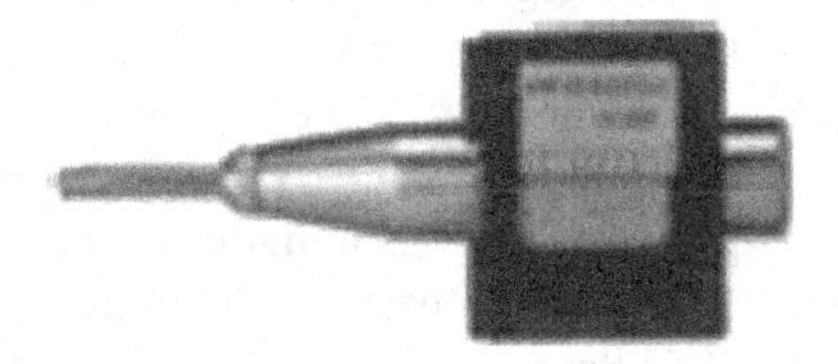

FIGURE 4.8 Type of rebound hammer.

FIGURE 4.9 Doing a test.

the calibration tools that come with the device when it is purchased. The calibration should be within the allowable limit based on the manufacturer's recommendation.

The first and the most important step in the test is to clean and smooth the concrete surface at the sites that will be tested by honing in an area of about 300 × 300 mm. Preferably, the surface should be tested to determine that it has not changed after

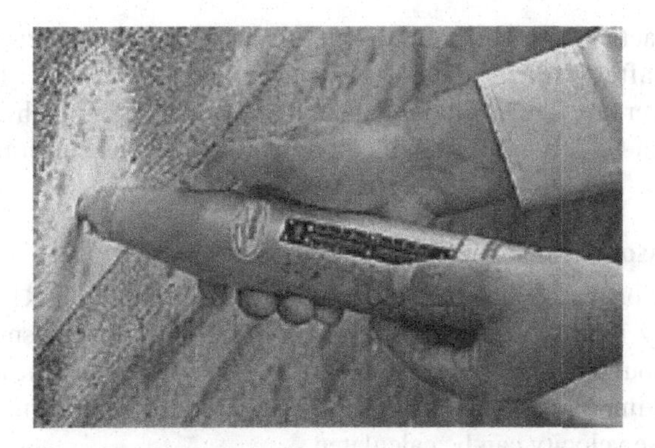

FIGURE 4.10 Test and read the number.

casting or that no smoothing took place during the casting process. On the surface to be tested, a net of perpendicular lines is drawn 2–5 cm apart in both directions. The intersection points will be the points to be tested; the test point must be about 2 cm from the edge. Figures 4.9 and 4.10 show that the surface must be cleaned before testing and that the rebound hammer is perpendicular to the surface.

The following recommendations should be followed during the test:

- The hammer must be perpendicular to the surface that will be tested in any conditions because the direction of the hammer affects the value of the rebound number as a result of the impact of hammer weight.
- The wet surface gives a significantly lower reading of the rebound hammer than a dry surface does—up to 20%.
- The tested concrete member must be fixed and not vibrate.
- The curves for the relationship between concrete compressive strength and rebound number as given from the manufacturer cannot be used; rather, the hammer must be calibrated by taking a reading on concrete cubes and crushing the concrete cubes to obtain the calibration of the curves. It is important to perform this calibration from time to time as the spring inside the rebound hammer loses some of its stiffness with time.

Only one hammer must be used when a comparison is made between the quality of concrete in different sites. The type of cement affects the readings, as in the case of concrete with high-alumina cement, which can yield higher results (about 100%) than those of concrete with ordinary portland cement. Concrete with sulfate-resistant cement can yield results about 50% lower than that using ordinary portland cement. A higher cement content gives a reading lower than that of concrete with lower cement content; in any case, the gross error is only 10%.

4.4.2.2.1 Data Analysis

The number of readings must be high enough to give reasonably accurate results. The minimum number of readings is 10, but usually 15 readings are taken. Extreme values will be excluded and the average taken for the remaining values. From this, the concrete compressive strength will be known.

The best practical method to enhance the accuracy of the reading is to pour out a cube and after 28 days take measurements using the hammer for 15 readings and take the average. Crushing the cube using the crushing machine and the difference can then be defined from which the hammer can be calibrated to increase its accuracy.

4.4.2.3 Ultrasonic Pulse Velocity

This test is one of the nondestructive testing types (ACI 228-89-R1 1989; BS 1881 1971a,b; Bungey 1993); its concept is to measure the speed of transmission of ultrasonic pulses through the construction member by measuring the time required for the transmission of impulses. When the distance between the sender and the receiver is known, the pulse velocity can be calculated.

These velocities are calibrated by knowing the concrete strength and its mechanical characteristics. These can then be used for any other type of concrete using the same procedure to identify the compressive strength, dynamic and static moduli of elasticity, and the Poisson ratio. The equipment must have the capability to record time for the tracks with lengths ranging from 100 to 3000 mm accurately +1%. The manufacturer should define how the equipment works in different temperatures and humidities. It must be available as a power transformer sender and the receiver of natural frequency vibrations between 20 and 150 kHz, bearing in mind that the frequency appropriate for most practical applications in the field of concrete is 50–60 kHz. There are different ways for transmission of a wave to occur as surface transmission, as shown in Figure 4.11a and b.

Semidirect transmission is clearly shown in Figure 4.12a and b. The third type of wave transmission is a direct transmission, as shown in Figure 4.13a and b. It is given with the UT (ultrasonic testing) equipment—two metal rods with lengths of 250 and 1000 mm. The first is used to determine the zero measurement and the second is used in the calibration. In both cases, each rod indicates the time of the passage of waves through it.

Hence, the ends of the rod are appropriately connected by the sender and the receiver. They measure the time for pulse transmission and compare it with the known reading; the smaller rod, if there are any deviations, adjusts the zero of the equipment to provide the known reading. A long bar is used in the same way to define accuracy of results; in this case, the difference between the two readings cannot be more than ±0.5% to qualify them to have the required accuracy.

The wave transmission velocity in steel is twice that in concrete; thus, performing a test on steel bars in a concrete member will influence the accuracy of the reading as the wave impulse velocity will be high. To avoid this, the location of the steel reinforcement must be defined previously with respect to the path of the ultrasonic pulse velocity.

To correct the reading, one must consider the reading for steel bars parallel to the path of the pulse wave, as shown in Figure 4.14. The calculation of the pulse velocity will be as given in the following equation. The time of transmission will be

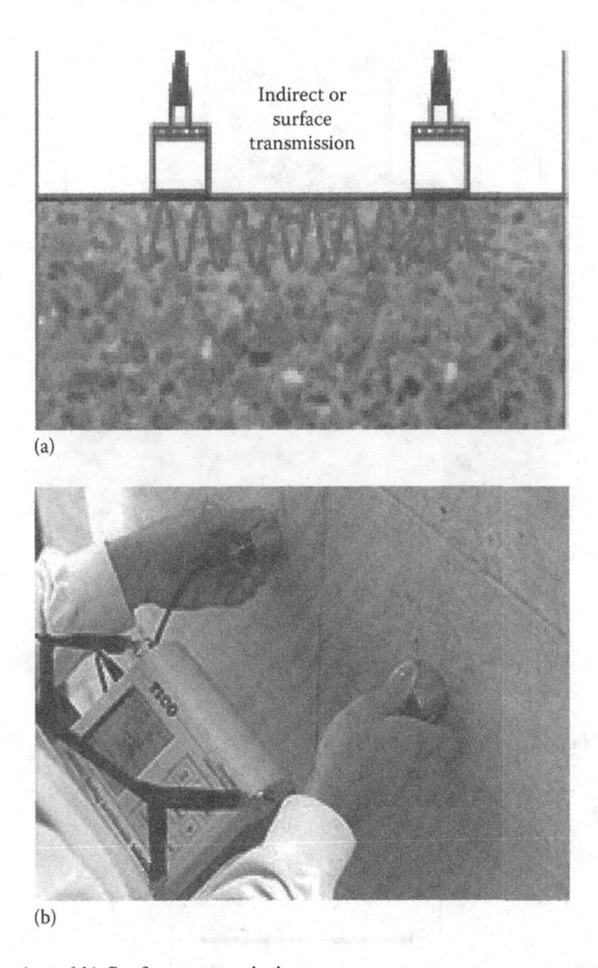

FIGURE 4.11 (a and b) Surface transmission.

FIGURE 4.12 (a and b) Semidirect transmission.

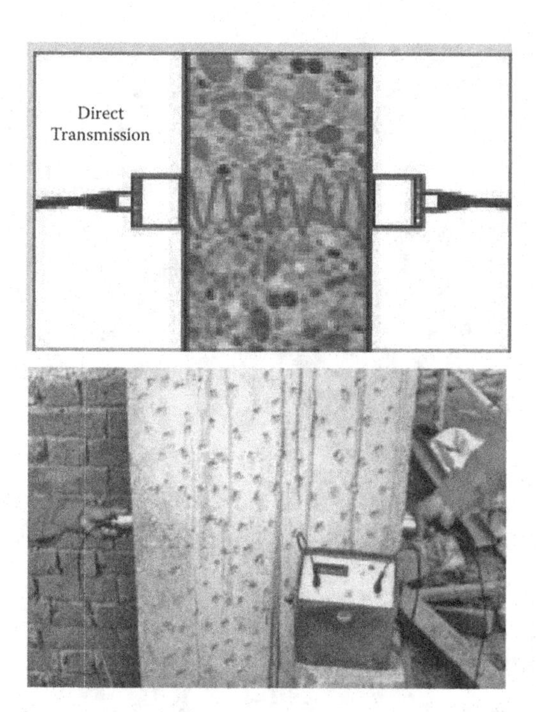

FIGURE 4.13 (a and b) Direct transmission.

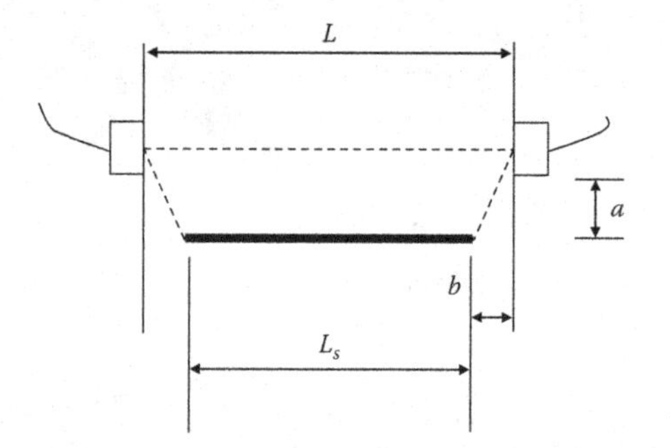

FIGURE 4.14 UT wave parallel to the steel bars.

calculated by the equipment and by ascertaining the wave path, the wave velocity in steel, and the distance between the two ends:

$$V_C = K \cdot V_m \tag{4.3}$$

$$K = \gamma + 2\left(\frac{a}{L}\right)\left(1-\gamma^2\right)^{0.5} \tag{4.4}$$

$$L_s = (L - 2b) \tag{4.5}$$

where

V_m is the pulse velocity from transmission time from the equipment
V_C is the pulse velocity in concrete
γ is a factor whose value varies according to steel bar diameter

The effect of the steel bar can be ignored if the diameter is 6mm or less or if the distance between the steel bar and the end of the equipment is large.

In the case of steel reinforcement bars, the axis is perpendicular to the direction of pulse transmission, as shown in Figure 4.15. Steel will have less effect on the reading in this case (Figure 4.16). The effect can be considered to be zero if a transmission source of 54kHz is used and the diameter of the steel bar is less than 20mm. The preceding equation is used in the case of smaller frequency, a diameter greater than 20mm, and by changing the value of γ according to the bar diameter by the data delivered by the equipment.

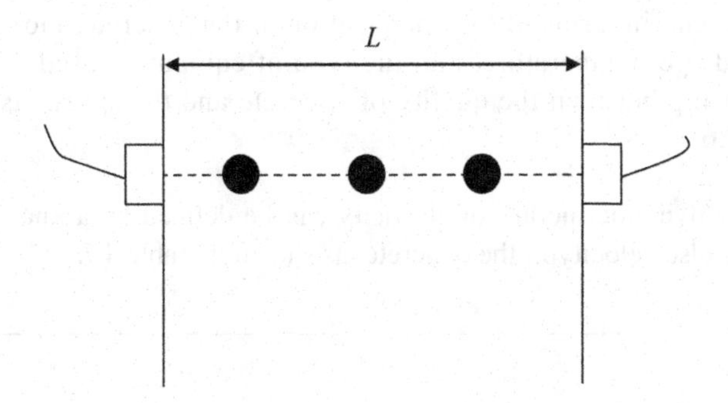

FIGURE 4.15 UT wave perpendicular to the steel bars.

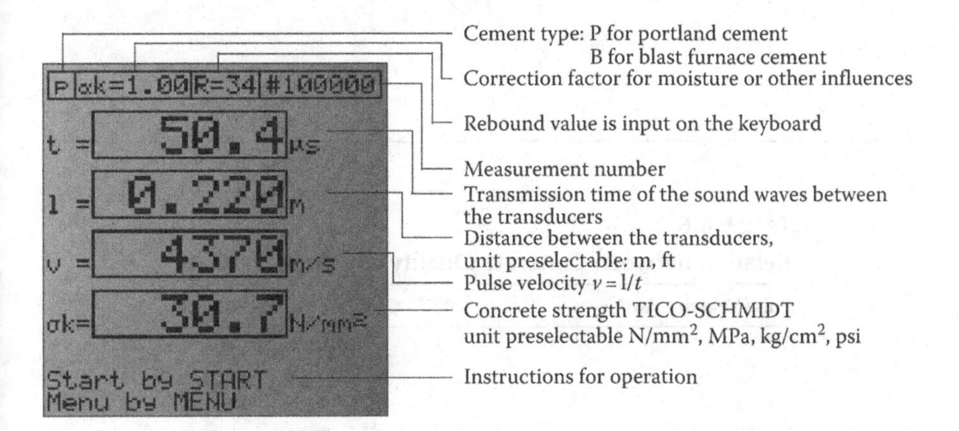

FIGURE 4.16 Ultrasonic pulse velocity screen.

Other factors influencing the measurements, namely, temperature, thawing, concrete humidity, and the factor effect must be considered. The *most common errors* include

- Ignoring the use of the reference bar to adjust zero, which will impact the accuracy of the results.
- The concrete surface is level. Smoothing it after pouring may result in properties different from the concrete in the core of the member, and therefore such measurements should be avoided as much as possible. If avoidance is not possible, one must take into account the impact of the surface.
- Temperature affects the transmission of ultrasonic velocity. According to Table 4.5, this fact must be taken into account for an increase or decrease in temperature of 30°C.
- When comparing the quality of concrete between the various components of the same structure, similar circumstances should be taken into account in terms of the composition of the concrete, moisture content and age, temperature, and the type of equipment used. There is a relationship between the quality of concrete and the speed, as given in Table 4.6.

The static and dynamic moduli of elasticity can be defined by a knowledge of the transmission pulse velocity in the concrete, as shown in Table 4.7.

TABLE 4.5
Temperature Effect on Pulse Transmission Velocity

Percentage Correction of the Velocity Reading (%)		Temperature (°C)
+4	+5	60
+1.7	+2	40
0	0	20
−1	−0.5	0
−7.5	−1.5	−4

TABLE 4.6
Relation between Concrete Quality and Pulse Velocity

Pulse Velocity (km/s)	Concrete Quality Degree
>4.5	Excellent
4.5–3.5	Good
3.5–3.0	Fair
3.0–2.0	Poor
<2.0	Very poor

TABLE 4.7
Relation between Elastic Modulus and Pulse Velocity

Elastic Modulus (MN/mm²)		Transmission Pulse
Static	Dynamic	Velocity (km/s)
13,000	24,000	3.6
15,000	26,000	3.8
18,000	29,000	4.0
22,000	32,000	4.2
27,000	36,000	4.4
34,000	42,000	4.6
43,000	49,000	4.8
52,000	58,000	5.0

4.4.2.4 Load Test for Concrete Members

This test is done under the following conditions:

- If the core test gives results of concrete compressive strength lower than results of characteristic concrete strength, which is defined in the design.
- If this test is included in the project specifications.
- If there is doubt about the ability of the concrete structure member to withstand design loads.

This test is usually performed on the slabs and, in some cases, the beams. This test exposes the concrete slab to a certain load and then removes the load; during this period, it measures the deformation on the concrete member as deflection or presence of cracks and compares it with the allowable limit in the specifications.

4.4.2.4.1 Test Preparation

This is done by loading the concrete member with a load equal to the following:

$$\text{Load} = 0.85(1.4\,\text{dead load} + 1.6\,\text{live load}) \tag{4.6}$$

The load is applied by using sacks of sand or concrete blocks. In the case of sand, sacks are calibrated to at least 10 sacks for every span, about 15 m² through the direct weight of the sacks. These sacks are chosen randomly to determine the weight of an average sack. The sacks are then placed on the concrete member that will be tested; the distance between vertical sacks needs to be considered to prevent an arching effect. As for the concrete blocks, their weight should be measured and they should be calibrated. In addition, the horizontal distance between them should be taken into account to avoid the influence of the arching effect.

It is important to identify the adjacent elements that have an impact on the structure element to be loaded in order to obtain the maximum possible deformation for the test member. Before load processing, the location of the test must be defined by identifying the places where the gauges will be placed, as well as calculating the

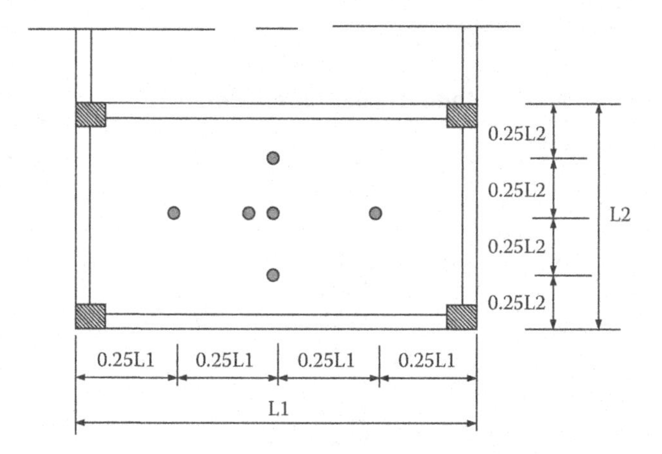

FIGURE 4.17 Location of measurement devices.

actual dead load on the concrete member, through the identification weight of the same member. In addition, coverage such as tiles, which will be installed on a slab of concrete, as well as lower coverage should be considered and plastering or weight of any kind of finishing work should be included. The location of the measurement unit is shown as an example for a slab test in Figure 4.17. The figure illustrates the following specifications:

- It places it in the middle of the span; placed beside it is another as a reserve.
- Another measurement device is placed at a quarter of the span from the support; the consultant engineer must define the other reasonable location for measurement.
- The measurement devices must be calibrated and certified before use—preferably with a smaller sensitivity of 0.01 mm; its scale is about 50mm.
- It is necessary to have devices that measure the crack widths; this device must have an accuracy of 0.01 mm.

4.4.2.4.2 Test Procedure

Define load test=0.85 (1.4 dead load+1.6 live load)—a dead load that has already affected the member.

1. Take the reading of the deflection before starting the test (R1).
2. Start placing the sack bags with 25% of the test load and avoid the arching effect or any impact load.
3. Read the measurement for the effect of 25% of the load and visually inspect the member to see if there are any cracks. If cracks are present, measure the crack width.
4. Repeat this procedure three times; each time, increase 25% of the load.
5. Record the time for the last load and the last deflection reading and crack thickness.

6. After 24 hours of load affects the time record, draw the location of the cracks, the maximum crack thickness, and the deflection reading (R2); then remove the load gradually and avoid any impact load.
7. After removing the entire load, measure the deflection reading and crack width.
8. Twenty-four hours after removing the load, record the measurement and the reading (R3) and record the crack width.

4.4.2.4.3 Resulting Calculations

The maximum deflection from load effect after 24 hours is as follows:

$$\text{Maximum deflection} = (\text{First measurement after passing 24 hours from load effect} - \text{reading before load effect}) \times \text{Device sensitivity} \quad (4.7)$$

$$\text{Maximum deflection} = (\text{R2} - \text{R1}) \times \text{Device sensitivity} \quad (4.8)$$

If any problems occur in the use of the first device, the second device reading should be used. If the reading of the two devices is close, the average of the two readings should be taken. The remaining maximum deflection after 24 hours of completely removing the load will be from the following equation:

$$\text{Maximum remaining deflection} = (\text{Reading of the device 24 hours after removing the load} - \text{Reading before load effect}) \times \text{Device sensitivity} \quad (4.9)$$

$$\text{Maximum remaining deflection} = (\text{R3} - \text{R2}) \times \text{Device sensitivity} \quad (4.10)$$

The maximum recovery deflection is calculated as follows:

$$\text{Maximum recovery deflection} = \text{Maximum deflection} - \text{Maximum remaining deflection} \quad (4.11)$$

The maximum deflection after 24 hours from loading and the maximum recovery deflection are as shown in Figure 4.17. The relation between the load and maximum deflection in cases of loading and uploading should be drawn. The maximum crack thickness is calculated after 24 hours from load effect and 24 hours after removing the load.

4.4.2.4.4 Acceptance and Refusing Limits

Calculate the maximum allowable deflection for the member as follows:

$$\text{The maximum allowable deflection} = L^2/2t, \text{cm} \quad (4.12)$$

where

L is the span of the member in meters—the shorter span in case of flat slab and short direction for solid slab and, in case of cantilever, twice the distance from the end of cantilever and the support face

t is the thickness of the concrete member in centimeters

Compare between the maximum deflection recorded after 24 hours from load effect and the allowable maximum deflection; there will be three outcomes:

1. If, after 24 hours, the maximum deflection from load effect is less than the allowable maximum deflection from the previous equation, then the test is successful and the member can carry the load safely.
2. If, after 24 hours, the maximum deflection from load effect is higher than the allowable maximum deflection, then the recovery deflection after 24 hours after removing the load must be equal to or higher than 75% of maximum deflection (recovery deflection ≥0.75 maximum deflection). If this condition is verified, then the member is considered to have passed the test.
3. If the recovery is less than 75% of the maximum deflection, the test should be repeated using the same procedure, but 72 hours after removing the load from the first test.

After repeating the test for a second time using the same procedure and precautions, this concrete structure member will be refused if not verified based on the following two conditions:

1. If the recovery deflection in the second test is less than 75% of the maximum deflection after 24 hours from load effect in the second test.
2. If the recorded maximum crack thickness is not allowed.

4.4.2.5 Comparison between Different Tests

The different methods to determine hardened concrete strength have advantages and disadvantages, which are summarized in Table 4.8. Moreover, these tests are different based on their costs, representation of the concrete member, and accuracy of the measured strength to the actual concrete strength; this comparison is illustrated in Table 4.9.

After the results of the tests are obtained, it will be found that the data on strength are lower than the concrete strength specified in the drawings or project specifications. The value of concrete strength in the drawing is the standard cube (cylinder)

TABLE 4.8
Comparison between Different Test Methods

Test Method	Probable Damage	Precaution Requirement
Overload test	Possible member loss	Member must be isolated or allowance of distributing the load to the adjacent members Extensive safety precaution in case of collapse failure
Cores	Holes to be made good	Limitations of core size and numbers Safety precaution for critical member
Ultrasonic	None	Need two smooth surfaces
Rebound hammer	None	Need a smooth surface

TABLE 4.9
Performance Comparison for Different Methods

Test Method	Damage to Concrete	Representative to Concrete	Accuracy	Speed of the Test	Cost
Overload test	Variable	Good	Good	Slow	High
Cores	Moderate	Moderate	Good	Slow	High
Ultrasonic	None	Good	Moderate	Fast	Low
Rebound hammer	Unlikely	Surface only	Poor	Fast	Very low

TABLE 4.10
Variation between Standard Cube Strength and Actual Strength On-Site

	Actual Strength/Standard Cube Strength after 28 Days	
Concrete Member	**Average**	**Range**
Column	0.65	0.55–0.75
Wall	0.65	0.45–0.95
Beam	0.75	0.6–1.0
Slab	0.5	0.4–0.6

compressive strength after 28 days. Note that the cube or the cylinder will be poured and compacted based on the standard; the curing process is for 28 days, so it is different from what will happen on the site. For example, the curing will not be for 28 days and the concrete slabs or columns cannot be immersed in water continuously for 28 days. The temperature on-site is surely different from and is more varied than that in the laboratory. In addition, the concrete pouring and compaction method for the cube or cylinder is different from that in the actual member size. All these variations were taken into consideration in the factors in the design codes, but the test measures the strength after complete hardening of the concrete. Table 4.9 shows a comparison between a standard cube test and *in situ* test for different reinforced concrete members (Table 4.10).

4.4.3 Sources of Concrete Failure

Corrosion is not the only source of failure. Many other sources cause deterioration of reinforced concrete structures; this must be kept in mind and understood well when an inspection is undertaken. These sources of failure include

1. Unsuitable materials
   a. Unsound aggregate.
   b. Reactive aggregate.
   c. Contaminated aggregate.

  d. Using the wrong type of cement.
  e. Cement manufacturer error.
  f. Wrong type of admixture.
  g. Substandard admixture.
  h. Contaminated admixture.
  i. Organically contaminated water.
  j. Chemically contaminated water.
  k. Wrong kind of reinforcement.
  l. Size error of steel bars.
2. Improper workmanship
  a. Faulty design.
  b. Incorrect concrete mixture (low or high cement content and incorrect admixture dosage).
  c. Unstable formwork.
  d. Misplaced reinforcement.
  e. Error in handling and placing concrete (segregation, bad placing, and inadequate compacting).
  f. Curing incomplete.
3. Environmental factors
  a. Soil alkali.
  b. Seawater or sewage.
  c. Acid industry.
  d. Freezing and thawing.
4. Structural factors
  a. Load exceeding design.
  b. Accident as blast load or dropped object.
  c. Earthquake load.

4.4.4 Example for Structure Evaluation

Our case study was an administration building located near the Red Sea in Egypt. The building was a steel structure constructed around 1975. It consisted of two stories; the ground floor was a room for employees waiting for their flights and the first floor was the administration office of an aviation and airport control room. The structure system was made of steel composed of beams and columns with reinforced concrete slab on steel corrugated sheet.

The Red Sea was only about 200 m away. The first impression in the visual inspection was that there had been no maintenance for a long time as the corrugated sheets were corroded in the lower parts outside the building for about 300 mm above the ground. The paint condition was generally good, but not as good as the original structural painting, so maintenance had been done periodically. During discussion with the owner (a very important step in any assessment), he mentioned that there had been no maintenance for around 5 years due to a shortage of funds. Near the bathroom, a sign of water could be seen on the false ceiling; therefore, it needed to be inspected in more detail. The false ceiling was

FIGURE 4.18 Corrosion in corrugated sheet.

removed and the corrugated sheet of the floor carrying the first floor was found, as shown in Figure 4.18.

The challenge was to define whether the corrugated sheet, which was corroded, was carrying a load or not. If it was working with concrete as a composite section, this meant that it was carrying a load and needed to be replaced. In this type of building, to carry out the repair by removing the slab and fixing a new composite section for the slab, it is necessary to hire a special contractor, which is expensive. Therefore, destroying the building and building a new one may be a better solution from an economic point of view. On the other hand, the corrugated sheet may be used as shuttering instead of wood and, after pouring concrete, kept as it is. Then, there would be no need to remove it. This method was famous in the 1970s; the specialized engineering and construction company is from the United States.

All the drawings and specifications do not exist, so it is necessary to collect information to confirm the structural system and whether either of these two methods of construction was used. If the slab is a composite section, the steel reinforcement will be light and always be a steel wire mesh for shrinkage. On the other hand, if it is only shuttering, it needs enough steel bars to carry the load, as it will be designed as a reinforced concrete slab.

Information was collected from the engineers who reviewed the design previously and others who supervised the building and a hole was made to see the steel bars; all these data confirm that there are steel bars of 13 mm and more, so this corrugated sheet is not needed to carry the load. Because the corrugated sheet is used as shuttering only, the building can exist longer.

In brief, an engineer should take his time in evaluation and not jump to a conclusion from first observation. He should not hesitate to collect data from any source; even if they do not add value, they may improve the data he already has. The decision is very critical; in this case, there are two different options: to destroy the building or to keep it for possibly another 30 years. Imagine the consequences if a wrong decision is made!

FIGURE 4.19 (a) General tank view, (b) cracks on the tank ring beam, (c) spalling of the concrete cover, and (d) reduction in steel cross-sectional area.

4.4.5 Example for Structure Assessment

There are cracks on the ring beam around the steel tanks. This tank is carrying oil mixed with a high percentage of water. The tanks are not located in a very bad environmental situation. They were constructed in about 1983; a photo is shown in Figure 4.19a. When an observation is made, a leak is found in the tanks that contain hydrocarbon and waters that affect the reinforced concrete beam. As shown in Figure 4.19b, the corrosion is parallel to the stirrups, every 200 mm and on the corner steel bars.

Therefore, the reason for and the shape of the cracks revealed that there is corrosion of the steel bars. In Figure 4.19c and d, some areas show a spalling of concrete cover and the reduction in the steel bar cross-sectional area is more than 20%; therefore, it needs complete repair. From a structural point of view, the steel bars have a tension force as the ring beam is always under tension. The repair procedure for this tank is described in detail in Chapter 6.

4.5 TEST METHODS FOR CORRODED STEEL IN CONCRETE

During the corrosion process, the volume of steel bars is increased and, due to volume increase, internal stresses form and cause the concrete to crack. These cracks may be vertical or horizontal, and they increase in length and width until the concrete cover falls. It is necessary to know the exact parts of concrete elements that cause a separation between concrete cover and concrete itself.

4.5.1 Manual Method

The deteriorated area can be defined by using a hammer to hit the concrete cover. If the sound that is made indicates air behind the surface of the concrete cover, the steel bars are corroded and concrete cover separation is a possibility. For large areas of concrete surface, such as concrete bridge decks, a steel chain can be moved on the surface to define the separated cover and the damaged parts. There are more complicated methods than using hammer or steel chain, such as using infrared or ultrasound.

Using the preceding inspection methods, an engineer can define the area that is expected to be corroded and predict concrete cover spalling after a short time; this must be considered in the next repair. A manual hammer or steel chain is used in this way on the surface deck of the bridge. This method is fast, easy to use, and less expensive than other methods such as radar, ultrasound, or infrared rays; this method can be used in big structures or when evaluating huge areas such as bridge decks.

The use of the hammer is often associated with the process of visual inspection. The hammer is easily carried, low cost, and quick to locate defects. It can make marks at the places where the sound indicates a vacuum behind the concrete surface. Use of infrared ray equipment is affected by the temperature; the best accuracy occurs when the temperature is as warm or as cold as possible. Through the camera in the infrared, the engineer can clarify where the layer of concrete that separated from the concrete has a temperature different from that of the main concrete part, as shown in Figure 4.20. Infrared is used on the surfaces of bridges in the early morning or in the evening, when the air is clean and pure, because then the area of separation between concrete cover and concrete can be identified.

Radar is also used in determining areas that have been separated from the concrete cover of a concrete structure; the United States uses a system of radar mounted on wheels to inspect bridge surfaces. Figure 4.21 illustrates radar fitted on a vehicle inspecting the bridge surface. In Europe and the United Kingdom, radar is activated

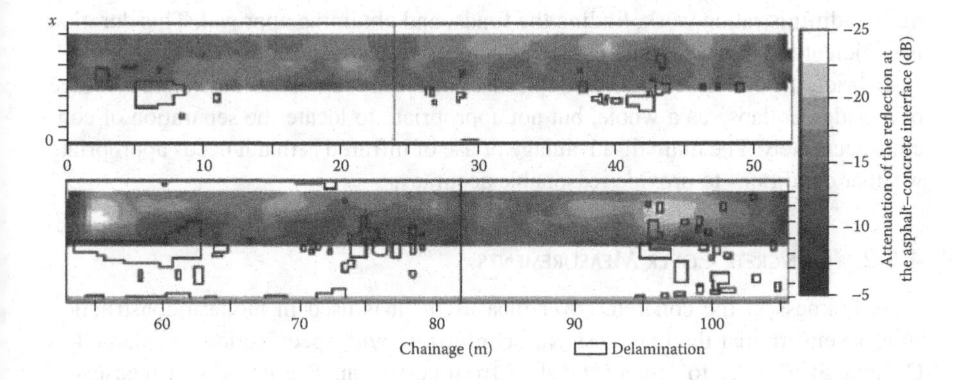

FIGURE 4.20 Measurement results of a bridge deck.

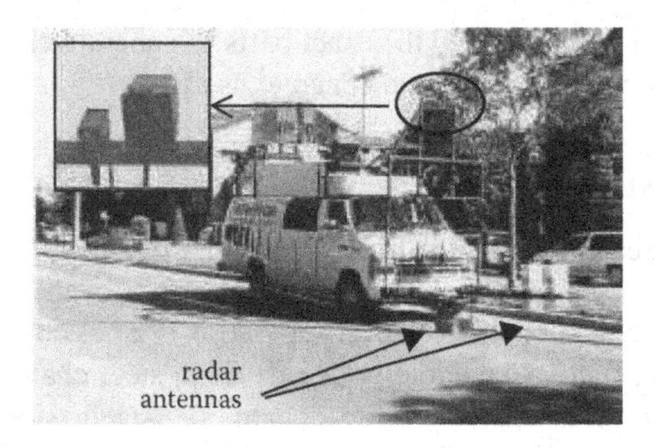

FIGURE 4.21 Radar antennas.

manually and used in buildings and different structures rather than on bridges. Many studies have been conducted using this equipment (Cady and Gannon 1992; Titman 1993). Note that radar is not accurate in defining the size and location of an area that has a separation on the concrete cover. Rather, it generally defines all the separated parts on the bridge as a whole (Alongi, Clemena, and Cady 1993). In the United States, radar is used together with infrared to increase the accuracy in defining a defective area, but this is expensive. Note also that infrared is used less frequently because that method needs reasonable temperatures during use.

In general, it is also worth noting that these methods, whether manual or using radar and infrared, need professionally trained workers, with high levels of experience and competence in reading the measurement and defining the defective area accurately. There are precautions and limits for the use of these methods: When water is found in the cracks, when the separation is deep inside the concrete, and when the steel bars are embedded with depth within the concrete, it will be difficult to define the separate areas as in structures like bridges that are always noisy due to moving cars. Therefore, a hammer cannot be used easily to define the defective area. Moreover, the existence of water and the depth of separation affect the readings in the case of radar and infrared.

It is worth mentioning that, in the execution of the repair process, an area of a size larger than that of the defective area defined before by the different methods should be removed, because there is time between structure evaluation and start of repair due to administration work, finding the funds, and obtaining approval. This duration is sufficient to increase the volume of concrete, which causes more separation.

In brief, the use of a radar is good in cases of estimating the volume on the surface of a bridge collapse as a whole, but not appropriate to locate the separation of concrete accurately. The main disadvantage of use of infrared is that it needs appropriate weather conditions to provide reasonable accuracy.

4.5.2 Concrete Cover Measurements

The thickness of the concrete cover measurement is used in modern construction so as to ensure that the thickness is in conformity with specifications (explained in Chapter 5) in order to protect the steel from corrosion. The process of measuring the thickness of the concrete cover in structures was marked by the beginning of the corrosion as the lack of concrete cover thickness increases the corrosion rate for corrosion as a result of chlorides or carbonation. This expedites the propagation inside the concrete, causing the speed of steel corrosion; also, the lack of cover helps in spreading moisture and oxygen, which form the main basis for the corrosion process.

The measurement of the concrete cover thickness explains the causes of corrosion and identifies areas that have the capability to corrode faster. This measurement (Alldred 1993) needs to be defined for *Y* and *X* axes in order to determine the thickness of the concrete cover at every point on the structure. The equipment that measures this thickness is simple and high-tech; measurement readings can be obtained as numbers. Figure 4.22 illustrates use of electromagnetic cover meter equipment on a bridge deck. Figure 4.23 presents equipment shape and the method of reading. Radiographs can be used for bridges, but the cost is high (Bungey 1993; Cady and Gannon 1992). The magnetic cover method is a simple method, but it is affected by the distance between steel bars. The thickness of the concrete cover exerts a large influence on the readings. This method depends on providing electricity with a 9 V battery and then measuring the potential voltage envelope by the device when current passes through the buried steel bars, as shown in Figure 4.24.

The British standard is the only standard concerned with measurement of concrete cover after construction (in Part 204 of BS 1881 1988b). In 1993, Alldred studied the accuracy of the measurement of the cover when there are more steel bars close to each other. He suggested using more than one head of measurement as that would increase the accuracy of the reading and the small heads have an impact on the accuracy of the equipment. Therefore, the problem with this method is that dense steel reinforcement in the concrete section will give inaccurate data; in this case, the equipment will be calibrated based on the existing steel bars as the reading is

FIGURE 4.22 Concrete cover measurement to a bridge deck.

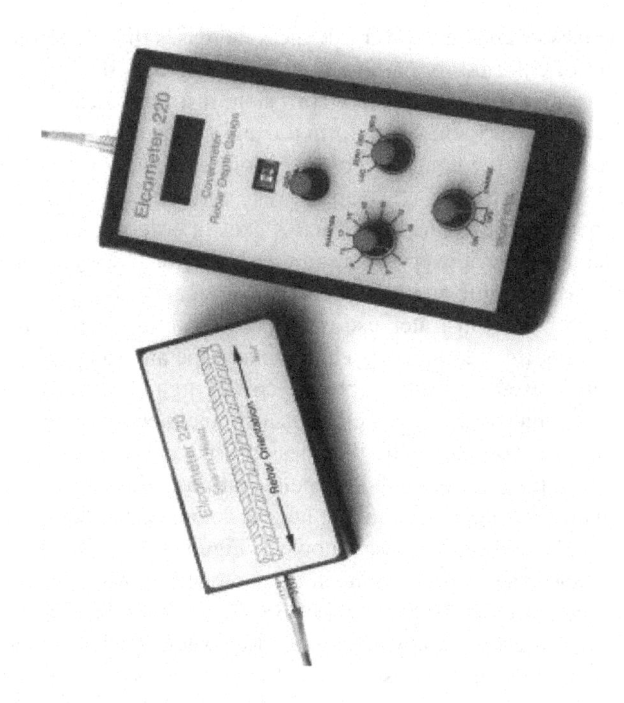

FIGURE 4.23 Concrete cover thickness measurement machine.

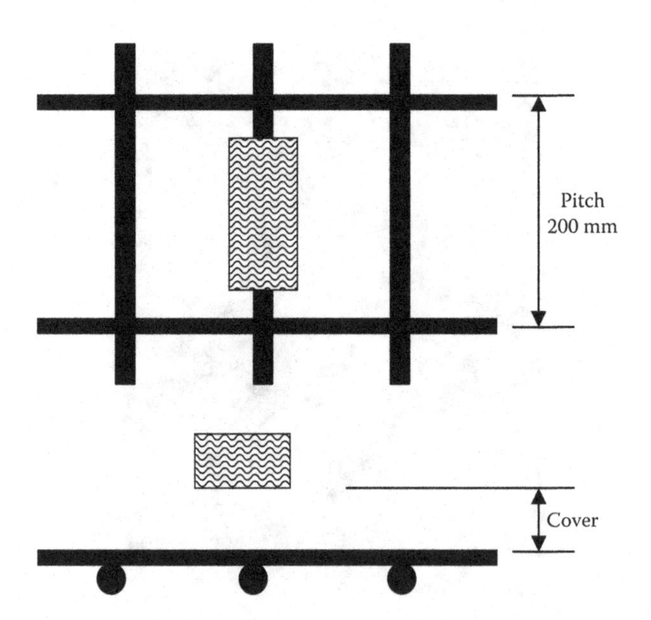

FIGURE 4.24 Device location on a concrete surface.

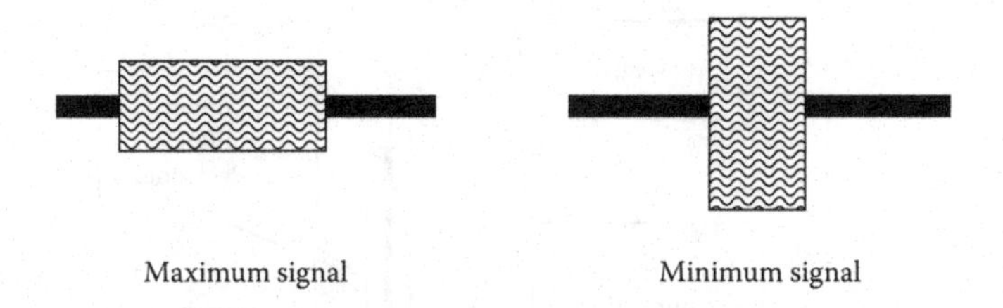

FIGURE 4.25 The device's right location.

FIGURE 4.26 Half-cell equipment.

affected by the type of steel. The person who works on the equipment must be competent and aware of anything that can affect the reading, such as bolts and steel wires (see Figures 4.25 and 4.26).

4.5.3 Half-Cell Potential Measurements

This system is used to determine bars that have lost passive protection. As explained in Chapter 2, the bars in the concrete that have lost passive protection were exposed to carbonating or existing chlorides, causing chemicals to react with the alkaline around the steel, which loses its passive layer. The half-cell equipment (ASTM C 876 1991) is composed of a rod of metal in solution by the same metal ions as a copper rod in saturated solution of copper sulfate. This is considered to be one of the most common types of equipment; some other pieces of equipment, such as bars of silver in silver chloride solution, can be linked through the voltmeter and other party potentiometers connected to the steel bars, as shown in Figure 4.27.

If the steel bars have a passive protection layer, the potential volts are small—0 to 200 mV copper/copper sulfate. If there has been a breakdown of the passive protection layer and some quantity of steel melted as a result of movement of ions, the potential voltage is about 350 mV. When the value is higher than this, the steel has already started to corrode. It is recommended to use half-cell equipment consisting of silver and silver chloride solution or mercury and mercury oxide solution. The equipment that consists of copper and copper sulfate solution is used but is not recommended because it requires regular maintenance and thus some readings will be erroneous.

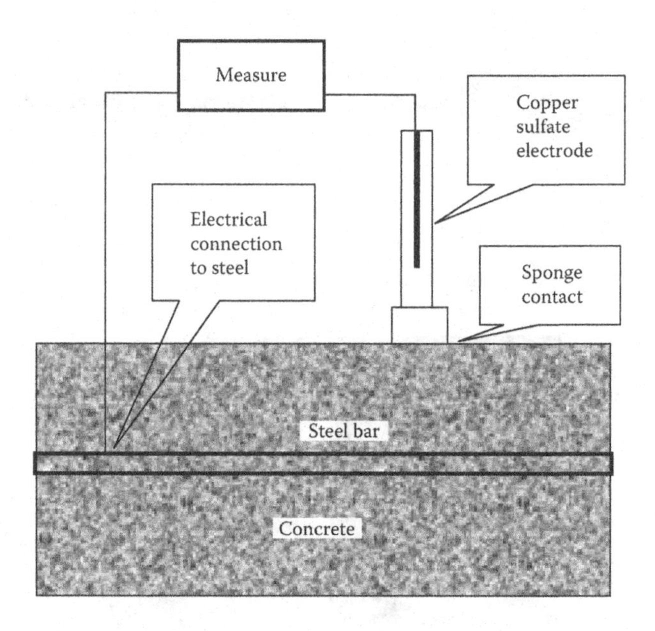

FIGURE 4.27 Drawing for half-cell device measurement.

To survey a large surface area of concrete to determine the bars that have been corroded, several half-cells on different locations on the surface need to be installed. These will be designated by the inspector. The gathering data will be represented as a contour line on the surface by which the location of the corroded bars underneath this large surface can be defined.

The measurement of the half-cell to the steel reinforcement bars is to define the probability of causing corrosion in the steel bars. The ASTM C867 specification presents a method to clarify the data obtained from measuring the potential voltage. These values are shown in Table 4.11 for copper/copper sulfate and silver/chloride silver equipment. The condition of corrosion is explained for each of the corresponding values.

TABLE 4.11
ASTM Specification for Steel Corrosion for Different Half-Cells

Silver/Silver Chloride	Copper/Copper Sulfate	Standard Hydrogen Electrode	Calomel	Corrosion Condition
Greater than −106 mV	Greater than −200 mV	Greater than +116 mV	Greater than −126 mV	Low (10% risk of corrosion)
−106 to −256 mV	−200 to −350 mV	+116 to −34 mV	−126 to −276 mV	Intermediate corrosion risk
Less than −256 mV	Less than −350 mV	Less than −34 mV	Less than −276 mV	High (<90% risk of corrosion)
Less than −406 mV	Less than −500 mV	Less than −184 mV	Less than −426 mV	Severe corrosion

The inaccuracy of such measurements is a result of the presence of water, which increases the negative values without corrosion occurring in steel, as wetting columns and walls give highly negative values for potential voltage regardless of the degree of steel corrosion. Negative voltage differs significantly underwater in offshore structures; in spite of the lack of oxygen, it reduces the rate of corrosion.

4.5.4 Electrical Resistivity Measurements

The process of steel corrosion includes electrical and chemical reactions; therefore, the electrical resistance and chemical composition of concrete are major factors that affect the rate of corrosion of steel. The corrosion rate in the steel reinforcement depends on the movement of ions from the anode to the cathode. A four-probe resistivity operation is used to measure the resistance of the soil to electricity; then, some changes are made to this system to measure the electrical resistance of concrete. This equipment is shown in Figure 4.28.

The resistance of concrete measured by the four-probe operation allows electric current to pass between two outside probes and measures the potential voltage between the two inner probes, as illustrated in Figure 4.29. From this, an engineer can measure the electrical resistivity of the concrete. Some equipment is needed to make a hole in the concrete to fix the probes, but the new equipment places it only on the surface.

The electrical resistance is greatly affected by the moisture in the internal porous concrete as well as by the quality of the content—that is, the *w/c* ratio in the concrete mix, curing, and using additives. There was a study performed by Hedjazi

FIGURE 4.28 Electrical resistivity measurement equipment.

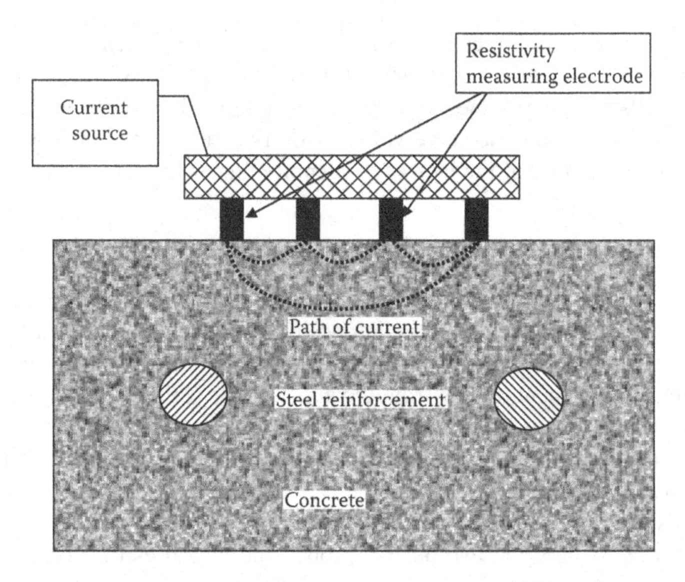

FIGURE 4.29 The workflow of electrical resistivity measurement equipment.

and Kabir (2022) to define the factors that affecting the electrical resistivity. In this study, surface electrical resistivity of concrete cylinders was measured from 3 to 161 days for concrete mixtures with four varying water-cement ratios (w/c) (0.45–0.60) and three distinct cement types. The study investigated the influence of important durability parameters such as cement type, long-term curing period, and w/c on concrete electrical resistivity. In addition, the impact of cylinder size on electrical resistivity of concrete was observed. The findings show that electrical resistivity of concrete decrease with increasing w/c, except for concrete with cement Type-I/II, which showed a minor increase in resistivity with a w/c of 0.55. Concrete with Type-V cement showed the highest electrical resistance. Note that the level of chloride does not have much impact on the electrical resistance of concrete.

During the measurement, there must be a greater distance between the probes of the largest size of aggregate; measuring the electrical resistance of only a piece of aggregate is avoided. It is possible to measure away from the steel bar, so the readings must be taken perpendicular to the steel reinforcement. Figure 4.29 illustrates an appropriate position for taking readings.

The measurement of the electrical resistance of the concrete cover occurs only through the use of two probes. The first probe is in the steel reinforcement and the second moves in the concrete surface. This is shown by the results obtained on the electrical resistance of concrete and can be distinguished through the comparison that follows, according to Broomfield et al. (1987; 1993):

>120 kΩ·cm: Low corrosion rate.
10–120 kΩ·cm: Low to medium corrosion rate.
5–10 kΩ·cm: High corrosion rate.
<5 kΩ·cm: Very high corrosion rate.

Many researchers have studied the relation between electrical resistivity and corrosion rate by using electron probes and have concluded the following:

>100 kΩ · cm: cannot differentiate between passive and active steel reinforcement.
50–100 kΩ · cm: slow corrosion rate.
10–50 kΩ · cm: from medium to high corrosion rate with active steel to corrosion.
<10 kΩ · cm: no impact of electrical resistivity on the corrosion process.

From the previous relation, this method is considered a reasonable way to determine the corrosion rate in steel reinforcement.

The electrical resistance is expressed on the corrosion rate of steel reinforcement and is also a good representative of the concrete quality control and density, which depends on the compaction method on-site. But, according to a study from Liverpool University, using four poles to measure the electrical resistance is incorrect because of the effect of steel on the reading. If the angle of measurement is perpendicular to the steel, the reading error can be reduced to the minimum. The study also found electrical resistance at more than 10 kΩ·cm. Active or passive protection in reinforced steel cannot be differentiated; if the electrical resistance is 10 km Ω·cm, then it is not effective in the process of corrosion.

4.5.5 Measurement of Carbonation Depth

Carbonation occurs in the depth by the propagation of carbon dioxide inside the cover of the concrete, changing the concrete alkalinity and reducing its value. The concrete alkalinity value is expressed by pH values equal to 12–13, with carbon dioxide propagation causing a reaction that reduces the alkalinity. This will affect the steel surface and then damage the passive protection layer around the steel bars; the corrosion process will then start.

Therefore, it is very important to define the depth of the concrete that transforms to carbonation and how far away from or close to the steel bars it is. This test (Parrott 1987) is performed by spraying the surface of broken concrete or breaking it by using special tools to obtain the carbonation depth by phenolphthalein dissolved in alcohol. This solution color becomes pink when it touches the surface of concrete with alkalinity value of pH of about 12–13.5, and the color turns gray or blue if concrete loses its alkalinity and pH value is less than 9. In this case, steel bars are losing the passive protection layer. In the last case, the alkalinity is lost and the pH becomes less than 9.

It is necessary that the part to be tested be newly broken whether it is a beam, a column, or a slab. After measuring the carbonation depth in the concrete cover and the distance from carbonation depth to the steel bar, whether it reaches the steel bars or not, it is easy to evaluate the corrosion risk for the steel reinforcement.

The best solution is to use phenolphthalein with alcohol and water: 1 mg phenolphthalein with 100 mm alcohol/water (50:50 mixing ratio) or more alcohol than water (Parrott 1987) if the concrete is completely dry, thus humidifying the surface by water. If the thickness of carbonation is about 5–10 mm, it breaks the passive layer to the steel bar 5 mm from the presence of changed color of the concrete surface.

Some of the aggregate or concrete mixtures have a dark color, making the reading of phenolphthalein very difficult; therefore, the surface should be clean when testing takes place. This test must be done in the areas accessible to the work of the breakers required in the concrete, as well as facilitate the work of repair for the broken part.

4.5.6 Chloride Tests

Chloride tests (AASHTO T260-84 1984; ASTM D 1411-82 1982) rely on an analysis of the samples of the concrete powder to determine the quantity of chloride. This test is done by using a drilling machine to rotary dig inside the concrete and extract concrete powder from the hole or through the broken part of the concrete. Several separate samples must be taken at different depths. The depth of the hole is varied to increase the accuracy and samples are always taken from 2 to 5 mm, first 5 mm from the surface. Often, the chloride concentration is very high, particularly in structures exposed to seawater and chlorides.

The holes are made by special devices that collect the concrete powder product that results from drilling the hole. The concrete powder will be taken at every drilling depth and will be added to a solution of acid, which is determined by the amount of chloride concentration. There are two principal ways of measuring *in situ* (quanta strips): the method to determine the electron–ion (specific ion electron) and the recent experience is more accurate but needs expensive equipment. Experiments of chloride concentration need trained professionals to conduct them.

When the readings of chloride concentration at various depths have been assembled, draw the shape to chloride concentration with depth from the surface of concrete into the concrete. From the general figure (Figure 4.30), it can be determined whether chlorides are within the concrete or from the impact of air and environmental factors affecting the concrete. Figure 4.30 shows the difference between the two cases.

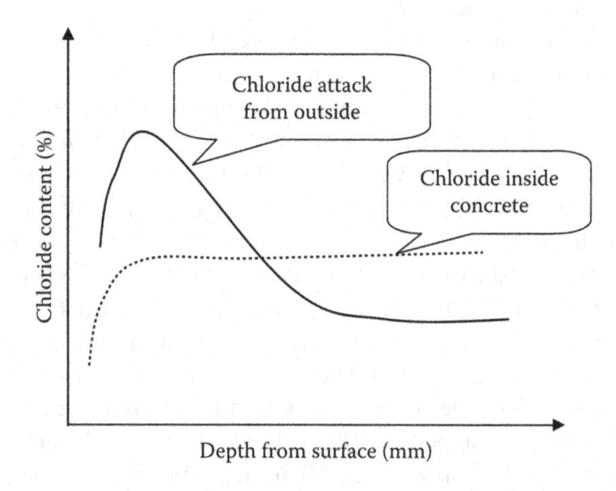

FIGURE 4.30 Comparison between a chloride attack from outside the concrete and inside the concrete.

From the previous figure, it can be seen that, in the case of the impact of chlorides from outside, the chloride concentration is high on the surface and then gradually decreases with depth until it reaches constant value. On the other hand, if the chloride exists inside the concrete, such as during concrete mixing due to use of salt water or an aggregate with high chloride content, the figure shows that near the surface the chloride concentration is minimal and increases with depth until it reaches constant values with the depth.

The chloride test method is done through melted acid and calibration work after that, based on specifications (BS 1881 1988a, Part 124). In addition to the method of melting, the melting acid can work in water; these tests follow ASTM D 1411-82 (1982) and AASHTO T260-84 (1984). The effect of chloride concentration on the steel reinforcement will be defined if the chloride concentration exceeds 0.6 from hydroxyl concentration, which causes failure in the passive layer around the steel reinforcement. This ratio is approximately 0.2%–0.4% chloride by weight of cement, 1 lb/yard^3 concrete, or 0.5% chloride by the weight of concrete. The figure for the distribution of chloride concentration with depth can determine the degree of impact of chloride in the steel reinforcement and the impact of chloride whether inside or outside the concrete.

4.6 BUILDING ASSESSMENT

The tests that have been detailed in the preceding sections are only tools to assist in the process of assessing a building. The main factor in the process—important to establishing the building assessment—is the experience of the engineer examining the structure. The evaluation of the building determines whether the structure is or is not subject to repair (e.g., because the situation is entirely bad or the cost of repair is near the cost of demolition and reconstruction).

The evaluation of the building is vital and critical to the decision process as decision-making will be based on an assessment of the building in terms of safety and of the repair cost. Therefore, the scope of repair work necessary for the building is based on the accuracy of the test results and the experience of the skilled labor force who conducts these tests. When the method of repair and its necessary cost are being determined, it is important to consider the methods required to protect the structure from corrosion after repair. Different ways of protection and their advantages and disadvantages will be illustrated in Chapter 7.

4.7 PREDICTING MATERIALS PROPERTIES

In ACI 562 provides a guideline of the material properties that can be used as in Table 4.12. If available drawings, specifications, or other documents do not provide sufficient information to characterize the material properties, it shall be permitted to determine such properties without physical testing from the historical data provided; however, the tools of inspection is not available so such in case if it is not possible to do the test or perform conceptual assessment you can use these tables considering that it can be used in USA, but you need to have these table for every country or depend on the experience of the engineer (Tables 4.13 and 4.14).

TABLE 4.12
Default Compressive Strength of Structure Concrete, psi

1900–1919	1000	2000	1500	1500	1000
1920–1949	1500	2000	2000	2000	2000
1950–1969	2500	3000	3000	3000	2500
1970–present	3000	3000	3000	3000	3000

Source: Adopted from ASCE/SEI41.

TABLE 4.13
Default Tensile and Yield Strength Properties for Steel Reinforcing Bars for Various Periods

		Structural	Intermediate	Hard				
	Grade	**33**	**40**	**50**	**60**	**65**	**70**	**75**
	Min. Yield Strength, psi	**33,000**	**40,000**	**50,000**	**60,000**	**65,000**	**70,000**	**75,000**
Year	**Min. Tensile Strength, psi**	**55,000**	**70,000**	**80,000**	**90,000**	**75,000**	**80,000**	**100,000**
1911–1959		X	X	X		X		
1959–1966		X	X	X	X	X	X	X
1966–1972			X	X	X	X	X	
1972–1974			X	X	X	X	X	
1974–1987			X	X	X	X	X	
1987-present			X	X	X	X	X	

Source: Adopted from SCE/SEI41.

X indicates the grade was available in those years.

The term structural, intermediate, and hard become obsolete in 1968, considering that hard grade does not correspond to metallurgical hardness.

4.8 ASSESSMENT CALCULATION

ACI 562-19 provides strength reduction values that will be used in case of the assessment, which is different from the rehabilitation design calculation that will be present in Chapter 8.

For the assessment, strength reduction factors are given in Table 4.15, which is higher rather than original design and repair. These increased values are justified by the improved reliability due to the use of accurate field-obtained material properties, and actual in-place dimensions and strength as per the tests that we discussed above.

The load factor for assessment and repair will be discussed in Chapter 8.

TABLE 4.14
Defult Tensile and Yield Strength Properties of Steel Reinforcement for Various ASTM Specification and Periods

				Structural	Intermediate	Hard				
			Grade	**33**	**40**	**50**	**60**	**65**	**70**	**75**
			Min Yield, psi	**33,000**	**40,000**	**50,000**	**60,000**	**65,000**	**70,000**	**75,000**
AASTM Designation	**Steel Type**	**Year Range**	**Mi. Tensile, psi**	**55,000**	**70,000**	**80,000**	**90,000**	**75,000**	**80,000**	**100,000**
A15	Billet	1911–1966		X	X	X				
A16	Rail1	1913–1966				X				
A61	Rail	1963–1966					X			
A160	Axle	1936–1964		X	X	X				
A160	Axle	1965–1966		X	X	X	X			
A185	WWF	1936–present						X		
A408	Billet	1968–1972		X	X	X				
A431	Billet	1974–1986								X
A432	Billet	1987-prest					X			
A497	WWF	1964–present							X	
A615	Billet	1968–1972			X					X
A615	Billet	1974–1986			X		X			
A615	Billet	1987–present			X		X			X
A616–96	Rail	1968–present					X			
A615	Axle	1968–present			X					
A615	Low alloy	1974–present					X		X	
A955	Stainless	1996–present			X		X			X

Source: ASCE/SEI41.

Rail bars are marked with letter R.

ASTM A706 has a min, tensile strength of 80 ksi but not less than 1.25 times the actual yield strength.

TABLE 4.15
Max. Strength Reduction Factor for Assessment

Strength	Classification	Transverse Reinforcement	ϕ
Flexural, axial or both	Tension controlled[a]		1.0
	Compression controlled[b]	spiral	0.90
		other	0.8
Sheaf, torsion, or both			0.80
Interface shear			0.80
Bearing on concrete[c]			0.80
Struts, ties, nodal, zone, and bearing areas in strut and ties model			0.80

[a] Applies when the steel tensile strain at member failure exceeds 2.5 εy where εy is the yield strain of the tensile reinforcement.

[b] Applies when the steel tensile strain at member failure does not exceed εy for sections in which the net tensile strain in the extreme tension steel at nominal strength is between the limits of compression-controlled and tension-controlled sections, linear interpolation of ϕ shall be permitted.

[c] Does not apply to post-tensioned anchorage zones or elements of structs and tie model.

REFERENCES

AASHTO T260-84. 1984. *Standard method of sampling and testing for total chloride ion in concrete ratio materials*. Washington, DC: American Association of State Highway Transportation Officers.

ACI 228-89-R1. 1989. *In-place methods for determination of strength of concrete*. Detroit, MI: American Concrete Institute.

ACI 562-19. 2019. *Code Requirements for Assessment, Repair, and Rehabilitation of Existing Concrete Structures*. Detroit, MI: American Concrete Institute.

Alldred, J. C. 1993. Quantifying the losses in cover-meter accuracy due to congestion of reinforcement. In Proceedings of the Fifth International Conference on Structural Faults and Repair, Vol. 2, pp. 125–130.

Alongi, A. A., G. G. Clemena, and P. Cady. 1993. Condition evaluation of concrete bridges relative to reinforcement corrosion. Vol. 3: Method for evaluating the condition of asphalt-covered decks. Strategic Highway Research Program, SHRP-S-325. Washington, DC: National Research Council.

American Concrete Society. 1987. *Concrete Core Testing for Strength. Technical Report No. 1, Including Addendum*, Detroit, MI: American Concrete Institute.

ASTM C 42-90m. 1990. *Standard Test Method for Obtaining Strength and Testing Drilled Cores and Sawed Beams of Concrete*. Philadelphia, PA: American Society for Testing and Materials.

ASTM C 876. 1991. *Standard Test Method for Half-Cell Potentials of Uncoated Reinforcing Steel in Concrete*. Philadelphia, PA: American Society for Testing and Materials.

ASTM D 1411-82. 1982. *Standard Test Methods for Water-Soluble Chlorides Present as Admixes in Graded Aggregate Road Mixes*. Philadelphia, PA: American Society for Testing and Materials.

Broomfield, J. P., P. E. Langford, and R. McAnoy. 1987. Cathodic protection for reinforced concrete: Its application to buildings and marine structures. In Proceedings of Corrosion/87 Symposium on Corrosion of Metals in Concrete, paper 142, NACE, Houston, TX, pp. 222–325.

Broomfield, J. P., J. Rodriguez, L. M. Ortega, and A. M. Garcia. 1993. Corrosion rate measurement and life prediction for reinforced concrete structures. In Proceedings of Structural Faults and Repair—93, Vol. 2, Edinburgh, Scotland, UK, pp. 155–164.

BS 1881. 1971a. *Testing Concrete. Part 5: Methods for Testing Hardened Concrete for Other Than Strength.* London, UK: British Standard Institution.

BS 1881. 1971b. *Testing Concrete. Part 6: Methods of Testing Concrete. Analysis of Hardened Concrete.* London, UK: British Standard Institution.

BS 1881. 1983. *Part 120. Method for Determination of the Compressive Strength of Concrete Cores.* London, UK: British Standard Institution.

BS 1881. 1988a. *Testing Concrete. Part 124: Methods for Analysis of Hardened Concrete.* Watford, UK: Building Research Establishment.

BS 1881. 1988b. *Testing Concrete. Part 204: Recommendations on the Use of the Electromagnetic Covermeter.* Watford, UK: Building Research Establishment.

Bungey, J. H. ed. 1993. Non-destructive testing in civil engineering. In International Conference of the British Institute of Non-Destructive Testing. Liverpool, England: Liverpool University.

Cady, P. D. and E. J. Gannon. 1992. Condition evaluation of concrete bridges relative to reinforcement corrosion. Vol. 1: State of the art of existing methods. SHRP-S-330. Washington, DC: National Research Council.

Concrete Society. 1984. Repair of concrete damaged by reinforcement corrosion. Technical report no. 26.

Hedjazi, S. and Kabir, E., 2002, Effects of cement type, water-cement ratio ratio, specimen size, and curing time on concrete electrical resistivity. *ACI Materials Journal* 119:22–23.

O'Reilly, M., J. Lafikes, O. Farshadfar, P. Vosough Grayli, O. Al-Qassag, and D. Darwin. 2022. Effect of concrete settlement cracks on corrosion initiation. *ACI Materials Journal 119*:117–124.

Parrott, L. J. 1987. *A Review of Carbonation in Reinforced Concrete: A Review Carried Out by C&CA under a BRE Contract.* Slough, UK: British Cement Association.

Titman, D. J. 1993. Fault detection in civil engineering structures using infrared thermography. In Proceedings of the Fifth International Conference on Structural Faults and Repair—93, Vol. 2, Edinburgh, Scotland, UK, pp. 137–140.

5 Codes and Specification Guides

5.1 INTRODUCTION

Here, we are going to discuss an issue of great importance as it directly affects the life of concrete structures: The design and construction specifications that should be followed in different codes to avoid corrosion of the steel reinforcement in concrete. If we want to construct a new building, we put in a lot of effort into solving problems of and paying money for construction. Thus, we must take extreme care to preserve the structure or we will find ourselves having to put more effort and money into maintaining it.

The proverb "an ounce of prevention is worth a pound of cure" fits the concept that codes and specifications provide us the means to control the corrosion of steel reinforcement in concrete. When a new building is constructed, it is important to follow all the precautions in design and construction on-site to prevent corrosion of steel in concrete. This is better than constructing a building without following any standards and then facing many problems in the future. This could mean going through costly repairs and, in some cases, not being able to return the building to its original strength.

Most different codes and technical specifications control concrete design and construction work to minimize the risk of corrosion in the steel reinforcements using the following ways:

- Define the maximum allowable limit of chloride in concrete.
- Define the components and mixture of the concrete cover.
- Define the concrete cover thickness.
- Define the maximum allowable crack width based on the design method.

Most specifications take into consideration the method of construction—for example, whether the concrete is cast *in situ,* is precast, or is prestressed—as well as the weather conditions that might affect the structure. Structures that are exposed to severe corrosion include marine structures that are directly exposed to seawater, such as ports, offshore platforms used in the oil industry, and bridge supports. The parts exposed periodically to dry cycles and cycles of seawater are especially exposed to corrosion. These parts make up the "splash zone" and are considered one of the most vulnerable elements of the structures, where the possibility of steel corrosion is very high.

One other type of structure that is frequently prone to corrosion is concrete bridges located in cold regions, such as those in Europe and some parts of North America. These countries use salt to melt the ice.

DOI: 10.1201/9781003407058-5

In the Middle East, in addition to buildings exposed to seawater, the climate has a very high relative humidity and high temperatures, which may cause corrosion of steel very quickly, as we explained in Chapter 3.

Therefore, the different specifications of each country have always been concentrated on that country's environmental conditions, as well as maintaining the quality control of concrete used in construction and the construction method commonly used in the country. Hence, it is clear that each country requires special precautions to be taken in the design, provided to the designer by the code, and specifications consistent with the weather and its environmental factors so as to avoid corrosion in steel reinforcement.

This chapter will illustrate the specifications used in American, Egyptian, and British codes. These codes have specifications concerning the percentage of chlorides in concrete and thickness of the concrete cover, as well as other specifications. These should be followed precisely to avoid the risk of steel corrosion.

5.2 ALLOWABLE CHLORIDE CONTENT IN CONCRETE

The majority of international specifications set maximum allowable amounts of chloride content in reinforced concrete because of its adverse impact on steel, as has been explained before. The sources of the chlorides in the concrete industry are often additives, some coarse aggregates or sand that may contain salts or the water used in the concrete mix. The percentage of chloride available in cement is very small and is controlled by the cement industry itself.

The specifications of the American Concrete Industry (ACI) place limits on the level of chlorides in reinforced concrete, as shown in Table 5.1. However, the American code shows different limits for the content of chlorides in concrete. It puts limits on the content of dissolved ions of chlorides (ACI 318R-89 1994) or total chloride ion content (ACI Committee 201 1994, ACI Committee 222 1994, and ACI Committee 357 1994) as a result of the components of concrete or due to chloride content resulting from chloride penetrating throughout the structure's lifetime. According to American specifications, ACI Committee 357 recommended that the water used in the concrete mix should not contain chloride amounts of more than 0.07% in the case of reinforced concrete, or 0.04% for prestressed concrete.

In the European Union code (European Prestandard ENV 206 1992), the limits for chloride content in concrete are identified by the type of application; the limits are 0.01% by the weight of cement in the case of plain concrete, 0.04% of the weight of cement in the case of reinforced concrete, and 0.02% of the weight of cement for prestressed concrete. These specifications prevent the use of any additives containing chlorides or using calcium chlorides in reinforced or prestressed concrete.

In the Egyptian code, the chloride ions in hardening concrete (present due to water, aggregate, cement, and additives) at the age of 28 days are not to exceed the limits, as shown in Table 5.2. The Egyptian code also recommends not using calcium

TABLE 5.1
Recommendation of Maximum Chloride Ion Content to Protect from Corrosion According to ACI Code

Structure Member Type	Maximum Chloride Content (% of Cement Weight)			
	Soluble in Water[a]	Total[b]	Soluble in Acid[c]	Total[4d]
Prestressed concrete	0.06		0.06	0.08
Reinforced concrete exposed to chloride in service	0.15	0.1	0.1	0.2
Reinforced concrete in dry conditions or protected from moisture in service	1.0			
Another fixed member	0.3	0.15		

[a] ACI 318R-89 (1994) ACI building code.
[b] ACI Committee 221.
[c] ACI Committee 357.
[d] ACI Committee 222.

TABLE 5.2
Maximum Soluble Ion Content to Protect from Steel Bar Corrosion

Maximum Soluble Chloride Ion Content in Water in Concrete (% of Cement Weight)	Surrounding Conditions
0.15	Reinforced concrete exposed to chlorides
1.0	Reinforced concrete dry and protected from humidity during usage
0.3	Other structure members

chloride additives or any additives with a base from chlorides for reinforced concrete, prestressed concrete, or concrete embedded with metal. The chloride ion content in the additives must not exceed 2% by weight of additives or 0.03% by weight of cement for reinforced concrete, prestressed concrete, or portland cement with sulfate-resistant cement.

These limits of chloride content in concrete can be achieved by testing the additives before use to ensure that the chloride is within limits and water that has salt in mixing or curing has not been used. Moreover, the percentage of chloride in the aggregate must be within the limits; the aggregate is usually washed by water before using it in the mix to clean it of salt and remove any other fine material.

5.3 CONCRETE COVER SPECIFICATIONS

The concrete cover is the first line of defense to protect the steel reinforcement from corrosion; therefore, complete attention should be paid to the thickness of the concrete cover. This thickness must not exceed a certain limit, according to the types of structure members and surrounding environmental conditions. Therefore, many codes have devoted specifications to the thickness of concrete cover according to the nature of the structure, method of construction, and quality of concrete used, as well as weather factors surrounding the structure.

5.3.1 British Standards

According to British standards, the ability of the concrete cover to protect the reinforcement steel bars from corrosion depends on the thickness of the concrete cover and the quality of the concrete. Moreover, when we talk about quality control of concrete—in particular, the concrete cover—we mean that the concrete has no cracks and good compaction since it will have a high density and a low water-to-cement (w/c) ratio. This prevents the ability of water to permeate easily into the concrete or any other material that will have an impact on steel reinforcement and cause corrosion.

Therefore, it is logical that concrete cover thickness is a function of the expected concrete quality; for example, in the case of high-quality concrete, there is little need for a thick cover. This philosophy is based on British standard BS 8110 (1985). The specifications of the concrete cover thickness in this British code are indicated in Table 5.3. This depends on the weather factors to which the concrete structure is exposed, as well as the concrete strength and its quality based on the cement content and w/c ratio.

5.3.2 American Codes

The specifications of the American ACI code do not give the exact details, as in the British code, but the least thickness of the concrete cover for concrete that has been poured at the site is shown in Table 5.4. To produce good concrete using water with a high salinity ratio, such as seawater, the American code sets the maximum w/c ratio equal to 0.4. In the case of structures exposed to seawater, the least allowable thickness of the concrete cover is 50 mm. Because of mistakes expected to occur during construction, it is preferred that the concrete cover design calls for a thickness of 65 mm so that, after execution, the minimum thickness of the concrete cover will be 50 mm. In the case of precast concrete, American code allows a concrete cover thickness that is lower than that shown in Table 5.4.

5.3.3 European Code

The European Union code gives precise and detailed recommendations and defines the degree of concrete strength required based on the weather conditions to which the structure is exposed. European code ENV 206 (1992) sets a water-to-cement ratio, as well as the lowest permissible content of the cement in concrete

TABLE 5.3
Properties and Thickness of Concrete Cover in BS 8110

Concrete Cover Thickness					
Concrete grade (MPa)	30	35	40	45	50
Water-to-cement ratio	0.65	0.65	0.55	0.50	0.45
Minimum cement content (kg/m³)	275	300	325	350	400
Environmental Conditions					
Moderate: Concrete surface protected from external weather or hard condition	25	20	20	20	20
Moderate: Concrete surface protected from rain or freezing and the concrete under water or the concrete adjacent to no affected soil		35	30	25	20
Hard: Concrete surface exposed to rain and wetting and dry			40	30	25
Very hard: Concrete exposed to seawater spray or melting ice by salt or freezing			50	40	30
Maximum condition: Concrete surface exposed to abrasion as seawater has solid particles, or moving water with pH 4.5 or machines or cars				60	50

TABLE 5.4
Minimum Cover Thickness for Cast-in-Place Concrete[a]

Minimum Cover (mm)	Type of Structure
75	Concrete deposited against the ground
Formed surfaces exposed to weather or in contact with ground	
50	No. 6 bar or greater
38	No. 5 bar or smaller
Formed surfaces not exposed to weather or in contact with ground	
38	Beams, girders, and columns
19	Slabs and walls, no. 11 bar or smaller
38	Slabs and walls, nos. 14 and 18 bars

[a] Per the ACI Committee 301 (1994).

and the least thickness of the concrete cover corresponding to the concrete strength according to weather conditions (Table 5.5). All choices will be made depending on the weather factors. This code has set specifications for reinforced concrete structures and prestressed concrete.

We must bear in mind that, during construction, the concrete cover thickness is often less than what the design calls for. Browne, Geoghegan, and Baker (1993) explored this topic and found the average thickness of the concrete cover to be about 13.9 mm, which is about half the value of that stated in the design (25 mm). Van Daveer (1975) has done a survey on the thickness of the concrete cover in the design

TABLE 5.5
Required Concrete Durability[a]

Exposure Condition	Maximum w/c Ratio	Minimum Cement Content (kg/m³)	Minimum Concrete Cover (mm)	Concrete Grade
Dry	0.65	260	15	C30/37
		Humid		
No frost	0.60	280	20	C30/37
Frost	0.55	280	25	C35/45
Deicing salts	0.50	300	40	C35/45
		Seawater		
No frost	0.55	300	40	C35/45
Frost	0.50	300	40	C35/45
		Aggressive Chemical		
Slightly	0.55	280	25	
Moderately	0.50	300	30	
Highly	0.45	300	40	

Note that the last column in Table 5.5, which determines the concrete strength characteristics, is set in megapascals (MPa); the first number is the value of the strength characteristics of cubes and the second number is the equivalent for the cylinders. For example, C30/37 means concrete characteristic cube compressive strength of 30 MPa and concrete characteristic cylinder compressive strength of 37 MPa. It is preferable to use cement resistance to sulfate sulfur if the sulfate content is more than 500 mg/kg in water or more than 3000 mg/kg in the soil. In both cases, additional painting on the concrete surface is recommended.

[a] Per European code ENV 206 (1992) and British specifications recommendation DDENV 206.

of bridges. If the design stated, for example, that the concrete cover thickness should be about 38 mm, it was found that the standard deviation of the thickness of the concrete cover was very high—up to 9.5 mm. As a result, in 1997 Arnon Bentur (Bentur and Jaegermann 1989, 1990) suggested that, to obtain a 50 mm thickness of concrete cover at a site, it was necessary to call for a cover thickness of about 70 mm in the design, the construction drawings, and the specifications.

The European code calculated the previous deviation occurring during construction, as the minimum concrete cover must increase by the allowable deviation. Its value ranges from 0 to 5 mm in cases of precast concrete and from 5 to 10 mm in cases of concrete cast *in situ*.

5.3.4 Specifications for Structures Exposed to Very Severe Conditions

Offshore structures are those that are directly exposed to seawater, such as ports or offshore platforms in the oil industry located in the sea or ocean. Based on the conditions, the ACI Committee 357 set certain specifications. The concrete cover thickness is defined based on whether reinforced concrete or prestressed concrete is used. In addition to that, limits of concrete strength and w/c ratio are stated in this specification, as shown in Table 5.6. The most dangerous area for corrosion in

TABLE 5.6
ACI Committee 357 Recommendation for Concrete Strength and Cover Thickness in Offshore Structures

Cover Thickness		Minimum Concrete	Maximum w/c	
Prestressed	Reinforced Steel	Strength at 28 Days	Ratio	Location
75	50	35	0.40	Air
90	65	35	0.40	Splash zone
75	50	35	0.45	Immersed in water

TABLE 5.7
Comparison of Different Specifications to Concrete Design in Splash Zone

Code	Concrete Cover Thickness (mm)	Maximum Crack Width (mm)	Maximum w/c Ratio	Minimum Cement Content (kg/m³)	Permeability Factor (m/s)
DNV	50		0.45	400	10–12
FIP	75	0.004×thickness or 0.3	0.45	400	
BS6235	75	0.004×thickness or 0.3	0.40	400	
ACI	65		0.40	360	

such structures is the splash zone, which is found in the region of the structure periodically exposed to seawater; here, it is not completely submerged in water, but not exposed only to air. There are several specifications for reinforced concrete in that particular region, as shown in Table 5.7.

The British code calls for a more detailed concrete cover thickness and concrete specifications required for private structures. Moreover, a correspondence between the degree of mixing and the quality of concrete and the chloride diffusion factor in the concrete was identified; this was also in accordance with the structure's life expectancy, as indicated in Table 5.8.

It is worth mentioning that the ability of the concrete cover to protect steel from corrosion depends not only on the thickness of the cover but also on the w/c ratio, the content of cement in the mix, and the quality control of concrete. While these factors are the most important influence, protection from corrosion also depends on the method of mixing, coarseness of the aggregate and sand sieve analysis, the way in which the concrete was compacted, and the curing of concrete after pouring.

5.3.5 Egyptian Code

The Egyptian code identified the thickness of the concrete cover required under the environmental conditions surrounding the structure. In this code, the concrete cover thickness depends on the surface of concrete that has tension stresses; the effects of

TABLE 5.8
British Code Requirement, Expected Diffusion Values, and Expected Time of Beginning of Corrosion,[a] as well as Corrosion Spread[b]

Source	Exposed Degree	Chloride Exposed Condition	Concrete Mix Detail: Cement Content (kg/m³)	Concrete Mix Detail: Maximum w/c Ratio	Concrete Mix Detail: Minimum Slump (mm)	Minimum Concrete Cover (mm)	Diffusion Coefficient (m²/s)	Age (Years): $C_i=0.4\%$	Age (Years): $C_p=1\%$
All structures BS 8110	Very severe	Spray seawater or deicing	325	0.55	40	50	3.93×10^{-12}	3.1	5.6
			350	0.50	45	40	3.18×10^{-12}	2.6	4.6
			400	0.45	50	30	2.57×10^{-12}	1.9	3.7
	Severe	Abrasion and seawater contain solids	350	0.50	45	60	3.18×10^{-12}	5.8	10.4
			400	0.45	50	50	2.57×10^{-12}	5.4	10.2
Bridges BS 5400 Part 4	Very severe	Deicing or seawater spray	360	N/A	40	50	3.93×10^{-12}	3.3	6.1
			330	0.45	50	40	2.57×10^{-12}	3.0	5.5
	Severe	Abrasion by seawater	360	N/A	40	50	3.93×10^{-12}	5.5	10.3
			330	0.45	50	40	2.57×10^{-12}	5.7	10.3
Offshore structure BS 6349 Part 1	Always submerged	Below sea level by 1 m	350	0.50	n/a	>50, prefer 75	3.18×10^{-12}	3.3	7.2
							3.18×10^{-12}	7.4	16.2
	Tidal/splash	Less than lowest level by 1 m	400	0.45	N/A	>50, prefer 75	2.57×10^{-12}	5.4	10.2
							2.57×10^{-12}	12.0	22.9
Seawater ENV06	XS1	Air saturated by water	330	0.50	40	35	3.93×10^{-12}	1.5	2.7
	XS2	Submerged in water	330	0.50	40	40	3.93×10^{-12}	2.0	3.6
	XS3	Spray or tidal	350	0.45	45	40	3.18×10^{-12}	2.1	4.6
Chlorides rather than seawater	XS4	Wet and rarely dry	300	0.55	40	40	3.93×10^{-12}	2.0	3.4
	XS5	Wet and dry cycles	330	0.50	40	40	3.93×10^{-12}	2.0	3.6

[a] $C_i=0.4$ by wt% of cement.
[b] Per unit of mass = 1% by weight of cement.

TABLE 5.9
Structural Elements Based on Environmental Conditions

Element Type	Degree of Exposed Tension Surface to Environmental Factors
Tension surface protected	All internal members in building
	The element is always immersed in water without any aggressive materials
	The last floor that isolated against humidity and rain
Tension surface not protected	All structures direct to weather as bridges and last floor is not isolated
	Structure's tension surface is protected but near coast
	Element exposed to humidity as cannot be away as open hall or parking
Tension surface exposed to aggressive environment	Element exposed to higher relative humidity
	Element exposed periodically to relative humidity
	Water tanks
	Structure exposed to vapors, gases, or chemicals with high effect
Tension surface exposed to oxides causing corrosion	Element exposed to chemical vapor causes corrosion
	Other tanks, sewage, and structure exposed to seawater

TABLE 5.10
Minimum Concrete Cover Thickness

	Concrete Cover Thickness (mm)			
	For All Elements Except Slab		Walls and Solid Slab	
Type of Element	$f_{cu} \leq 250$	$f_{cu} > 250$	$f_{cu} \leq 250$	$f_{cu} > 250$
Tension surface protected	20	15	15	10
Tension surface not protected	25	20	20	15
Tension surface exposed to aggressive environment	30	25	25	20
Tension surface exposed to oxides causing corrosion	40	35	35	30

environmental factors have been divided into four sections, as shown in Table 5.9. From this table, one can accurately determine the structure under any type of element. By knowing the strength characteristics of the concrete construction and the type of element, one can determine the thickness of the concrete cover, as shown in Table 5.10. The Egyptian code states that, in any case, thickness of the concrete cover should not be less than the largest bar diameter used.

5.3.6 Executing Concrete Cover

From the previous sections, the importance of concrete cover has become clear. Several practical methods are used to make sure that the thickness of concrete cover matches design requirements during the construction process. The most famous method uses a piece of concrete cube, called a "biscuit," of about 50×100mm dimension; the thickness depends on the required cover thickness, as shown in the photo in Figure 5.1. While pouring, a steel wire is inserted into this concrete cube and the steel bars are tied with these pieces to maintain the spacing between the bars.

The disadvantage of this method is that moving the workers on the steel bars to pour concrete or perform any other activities, such as inspection or supervision, exerts a concentrated pressure on the concrete pieces and can lead to cracks and damage. Therefore, after construction, in many cases, the cover thickness will be found to be very small or will vanish. The advantage of this method is that it is very inexpensive because the poured concrete is the same as that used on-site and the workers are also available on-site. This method can be improved upon by using the same concrete mix with the same quality control and by curing these pieces and using them after 14days after their full capacity to carry the load is reached, which prevents failure during the concrete pouring process.

A more practical method is to use a plastic piece that will maintain the concrete cover; it is cheaper and verifies that the thickness of concrete cover is as specified in the drawings. Recently, many contractors have been using these plastic parts, which are cheap and strong, to maintain the thickness of the cover with high accuracy. The photos in Figures 5.1 and 5.2 show different types of plastic pieces based on the bar diameter and the location of the steel bars. These plastic forms are used to stabilize the steel bars, thus maintaining the thickness of the concrete cover along the structural member during the construction of slab, beams, columns, or foundations.

Figure 5.1 shows the different types and forms of plastic cover, which vary according to the size and form of the bars and the concrete cover thickness to be preserved.

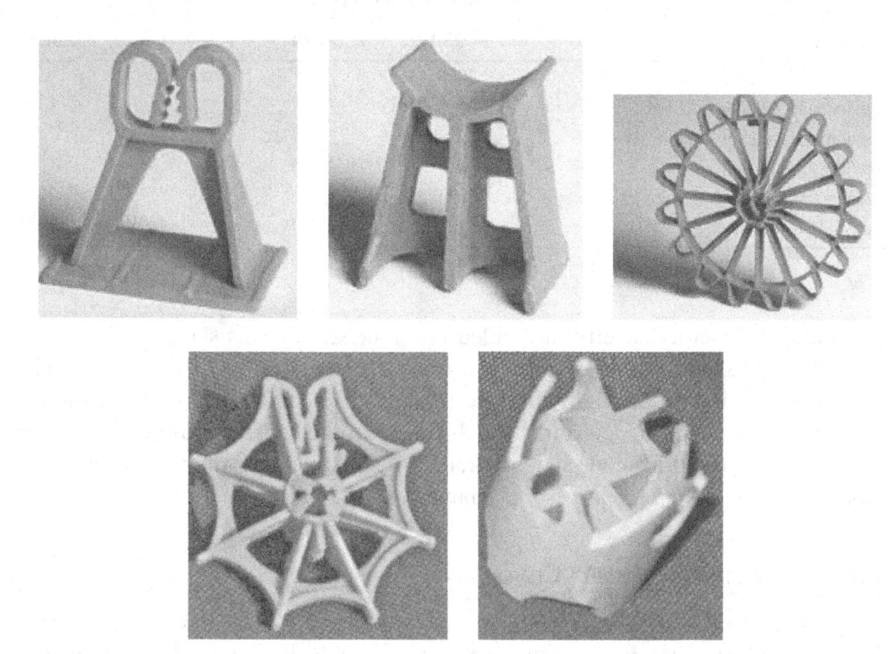

FIGURE 5.1 Photos of different kinds of plastic pieces.

FIGURE 5.2 Plastic pieces carrying chairs.

Therefore, the customer must give the company's suppliers all the required information on the quantity and thickness of the concrete cover and diameters of the steel bars to be used, as well as the type of concrete member that will be installed. There are different types of these pieces of plastic, depending on whether they are used in a column, slab, or beam. Recalling the picture of the biscuit form, it is worth mentioning that we have seen that this kind of "biscuit" can fix the steel bars and prevent them from moving, which also ensures that the distance between steel bars is strictly enforced. Figure 5.2 shows the plastic pieces carrying the chair; this is also useful for protecting any piece of steel that is exposed to the concrete's outer surface. In some cases, specifically for the foundation, we use a plastic pipe with its diameter as the required concrete cover with the advantage that it keeps the spacing like a plastic spacer and the concrete will fill the pipe so it will be very strong.

For more than 30 years, another method was used to keep the concrete cover from corrosion by placing a reasonable aggregate underneath the steel bars; however, this is now considered a very bad method and is prevented. Nevertheless, this method is still used in some low-cost residential buildings in developing countries because it is very inexpensive. However, future expenses will be incurred due to the deterioration of the building and the necessity for performing repairs.

After construction, it is recommended that specifications highlight the importance of measuring the concrete thickness for some special structures exposed to severe conditions. The British standard requires that the measurement of concrete cover be carried out in accordance with BS 1881 Part 204, using an electromagnetic device that estimates the position depth and size of the reinforcement. The engineer advises on locations for checking the cover and the spacing between measurements based on the objective of the investigation.

Reinforcement should be secured against displacement outside the specified limits unless specified otherwise. The actual concrete cover thickness should not be less than the required nominal cover minus 5 mm. Where reinforcement is located in

FIGURE 5.3 Electromagnetic cover meter.

relation to only one face of a member (e.g., a straight bar in a slab), the actual concrete cover should be not more than the required nominal cover plus:

- 5 mm on bars up to and including 12 mm in size.
- 10 mm on bars over 12 and up to and including 25 mm in size.
- 15 mm on bars over 25 mm in size.

A nominal cover should be specified for all steel reinforcement, including links. Spacers between the links (or the bars where no links exist) and the formwork should be of the same size as the nominal cover. Spacers, chairs, and other supports detailed on drawings, together with other supports as may be necessary, should be used to maintain the specified nominal cover to the steel reinforcement. Spacers and chairs should be placed in accordance with the requirements of BS 7973-2 and should conform to BS 7973-1. Concrete spacer blocks made on the construction site should not be used.

The position of reinforcement should be checked before and during concreting; particular attention should be directed at ensuring that the nominal cover is maintained within the given limits, especially in the case of cantilever sections. The importance of cover in relation to durability justifies the regular use of a cover meter to check the position of the reinforcement in the hardened concrete. Figure 5.3 shows the electromagnetic cover meter equipment, which is discussed in detail in Chapter 4.

5.4 MAXIMUM CRACK WIDTH

Cracks in the concrete cover have a direct impact on the process of corrosion in steel because the presence of cracks assists in the spread of carbon dioxide or chloride ions within the concrete. This triggers corrosion and is also the main reason for the entry of oxygen, thereby accelerating the process of corrosion and increasing the corrosion rate. Therefore, the ACI Committee 224 (1980) sets a maximum permissible width of cracks according to the environmental conditions surrounding the structure, as shown in Table 5.11.

When a concrete section is designed to resist corrosion, there are specifications that give the equations of the expected crack width due to load and stresses on the

TABLE 5.11
Relation of Environmental Condition and Allowable Maximum Crack Width

Environmental Condition	Maximum Crack Width (mm)
Dry air or protected surface	0.4
High humidity, water vapor, and soil	0.3
Seawater or sprayed seawater	0.15
Barrier walls to water except pipes under no pressure	0.1

structure. The ACI 224 Committee proposed the following equation to calculate maximum crack width:

$$W_{max} = 0.076\beta f_s (d_c A)^{0.333} 10^{-3} \tag{5.1}$$

where

W_{max} is the crack width in inches
β is the ratio of the distances to the neutral axis from the extreme tension fiber and from the centroid of the reinforcement
f_s is the calculated stress (ksi) in reinforcement at service load computed as the unfactored moment divided by the product of steel area and internal moment arm
d_c is the concrete cover in inches
A is the area of concrete symmetrical with the reinforcing steel divided by the number of the bars

According to clause 10.6.4 of the ACI code, satisfactory flexural cracks in beams and one-way slabs are achieved if the factor z, defined in Equation 5.3, does not exceed the following specified values. In arriving at these values, maximum crack widths of 0.33 and 0.41 mm for exterior and interior exposures, respectively, have been stipulated:

$$\begin{aligned} &\text{For exterior exposure}, z < 25.4\,\text{kN/mm} \\ &\text{For interior exposure}, z < 30.6\,\text{kN/mm} \end{aligned} \tag{5.2}$$

This requirement should be satisfied at sections of maximum positive and maximum negative bending moments and is only necessary if the steel used has a yield stress $fy > 280$ N/mm^2:

$$z = f_s \sqrt[3]{d_c A} \tag{5.3}$$

where

f_s is the calculated stress (N/mm^2) in the reinforcement bars at service loads, which may be determined by dividing the applied bending moment by the product of the steel area and the internal moment arm or may be considered to be $0.6fy$

d_c is the thickness of concrete cover (mm) measured from the extreme tension fiber to the center of the closest reinforcement bars

A is the effective tension area of concrete (mm^2) surrounding the flexural tension reinforcement and having the same centroid as the area of reinforcement, divided by the number of bars

If different bar sizes are used, the number of bars to be considered is obtained by dividing the total area of reinforcement by the area of the largest bars.

5.4.1 Recommended Reinforcement Details of Crack Control

Studying Equation 5.3 reveals that the following reinforcement detailing assists in reducing the value of z and hence leads to the development of narrower cracks:

- The use of the thinnest concrete cover for reinforcement bars consistent with the steel protection requirement and exposure conditions.
- Taking into consideration placement of larger-size bars in the row closest to the extreme tension fiber whenever more than a single row of reinforcement and different bar sizes are used in construction, procedure, and design detailing.
- Preferably using relatively larger numbers of small bar diameters rather than a small number of large bar diameters, given the same area of steel.
- The use of least practical distance between rows of bars (to allow the concrete to fill the forms, this distance should not be less than one bar size or 25 mm).

Note that the cracks occur not only due to loads but also as a result of exposure of the structure to the cycles of freezing and drying that cause plastic cracks or shrinkage cracks or that may occur as a result of some alkaline element in the aggregate. The role of cracks in the corrosion of reinforcement is controversial. Much research has shown that corrosion is not clearly correlated with surface crack widths in the range normally found with reinforcement stresses at service load levels. For this reason, the former distinction between interior and exterior exposure has been eliminated. These factors can be ignored by taking care regarding the components of concrete and also in the design of concrete mix, although it is difficult to forecast the crack width in such cases. In calculating crack width due to loads, the width of the crack as well as the direction of the crack is impressive.

Cracks in the direction of the main steel tend to be more influential than in the case of direction with secondary steel, where most of the steel will exhibit corrosion.

Beeby (1979) studied cracks and found that they are often in the direction of secondary steel. He notes that secondary steel does not cause important changes such as the main steel in that it does not bear any loads, which would make it dangerous for the safety of the structure and does not play a key role in carrying loads. However, Beeby has also mentioned that if the cover of the concrete falls as a result of the corrosion, the main steel will be corroded quickly; therefore, it is important that secondary steel is protected and also that the stirrups receive the same protection as the main steel does.

5.5 DESIGN PRECAUTIONS IN CASE OF CARBONATION

Carbonation of concrete can be measured by calculating rates of the spread of carbon dioxide in the concrete, as explained in Chapter 3. These equations can be used to anticipate the time required for carbonation to reach to steel. Time, t_i, is the time required until the beginning of corrosion. In addition, t_p is the time required for the propagation of corrosion in the steel prior to the spalling of the concrete cover. Estimating the value $t_i + t_p$ is the basis for controlling the carbonation, which causes corrosion.

Parrot (1994) conducted a great deal of research that showed the t_i and t_p clearly. Through some of the factors that are calculated in practice, including factors k, n, and often factor n (equal to 2), the coefficient k changes, depending on the types of concrete and environmental factors affecting the exposed structure. The value of the coefficient k can be identified through practice tests in normal weather. Some experiments were carried out in 1990 by Bentur and Jaegermann, who calculated t_i using

$$t_i = \left(\frac{d_c}{K}\right)^{0.5} \tag{5.4}$$

where

K is the factor for certain environmental factors and concrete with specific quality
d_c is the thickness of the concrete cover

With respect to the factor t_i, Parrot (1994) derived equations based on the following premises:

- The CO_2 factor is the diffusion of air into the concrete cover.
- Spreading of CO_2 in concrete depends on the CaO content.
- Change of the CO_2 content in air will be neglected.
- In cases of higher humidity, the $n = 2$ factor will be different than in the previous equation.

TABLE 5.12
Values m, n, Cement Content (c),[a] CR,[b] and tp[c]

Relative Humidity (%)	40	50	60	70	80	90	95	98	100
m	1.00	1.00	1.00	0.797	0.564	0.301	0.160	0.071	0.010
n	0.48	0.512	0.512	0.480	0.415	0.317	0.256	0.216	0.187
c (*CEM1b*)	460	460	460	460	485	535	570	595	610
c (*CEM2b*)	360	360	360	360	380	420	445	465	480
c (*CEM3b*)	340	340	340	340	355	395	420	440	450
c (*CEM4b*)	230	230	230	230	240	265	285	295	305
CR	0.3	0.3	0.3	2	5	10	20	50	10
t_p	330	330	330	50	20	10	5	2	10

[a] Kilograms per cubic meter.
[b] Micrometers per year.
[c] Year.

From these assumptions, Parrot proposed the following equation:

$$D = \frac{aK^{0.4} t_i^n}{C^{0.5}} \tag{5.5}$$

where

D is a depth of the transformation of carbon
K is the permeability of air (in units of 10^{-16}m^2) of the concrete cover

This value depends on relative humidity and can be estimated from the coefficient value of *K* at 60% relative humidity in $K = m\ K60$. The values of *m* with a different relative humidity are in Table 5.12.

The factor *n* represents the value of the higher power, always takes a value equal to 0.52, and decreases with increase in the relative humidity by almost more than 70% (Table 5.12).

The factor *c* is the content of CaO in the concrete cover. This factor depends on the local production of cement (as in Table 5.12) because it will have different values depending on relative humidity and the types of cement used in Europe.

The factor *a* is usually equal to 64, based on experimental results.

The factor t_p is present at the time of corrosion propagation on the steel and is calculated based on Parrot (1994) in the following equation:

$$t_p = \frac{CD}{CR} \tag{5.6}$$

where

CD is the corrosion depth in the steel reinforcement that causes cracks in the concrete cover

CR is the steel corrosion rate in the propagation period

The estimated value of *CD* by 0.1 mm has been explained before. The *CR* could account for value according to the relative humidity, as shown in Table 5.12. These approximate values have been identified based on information and process greatly affected by any external factors. The values in the table may vary by about 10%.

Based on the previous equation, the relation between the concrete cover thickness and the structure's lifetime ($t = t_i + t_p$) can be calculated as follows:

$$D = \frac{64K^{0.4}\left[t-(100/CR)\right]^{n}}{C^{0.5}} \tag{5.7}$$

Using this equation, by default, Parrot calculated the relationship between permeability of air and efficiency of the concrete strength by the quality of concrete (concrete strength and curing) and the thickness of the concrete cover to determine true lifetime of the structure.

TABLE 5.13
Required Concrete Quality under Different Weather Conditions[a] at Lifetime of 75 Years

Weather Condition	Cover Thickness	Minimum Cover Thickness	K60 (10^{-16}m²)	K (10^{-16}m²)	Concrete Cube Strength (MPa)	w/c Ratio	Relative Humidity
Dry	25	15	36	42	16	0.82	50
	25	15			20	0.90	
Moderate	25	15	46	42	15	0.85	65
	25	15	1.1	0.84	33	0.52	70
	25	15			25	0.65	
Dry/wet	30	20	1.7	1.2	31	0.55	50–100
	30	20	2.5	1.4	29	0.57	60–100
	30	20			35	0.55	
Wet	30	20	13	3.8	21	0.70	90
	30	20	46	7.4	15	0.85	95
	30	20			30	0.60	

[a] Calculated from Equation 5.3.

Considering the relationship between concrete strength after 28 days and time f_t after curing for more than 28 days, the air permeability of $K60$ can be calculated by the following equations:

$$f_t = f_{28}\left[0.25 + 0.225 log(t)\right] \quad (5.8)$$

$$log(K60) = \frac{(30 - f_t)}{10} \quad (5.9)$$

where t is the curing time plus 8 days.

The results of the calculations that have been inferred from the previous equation are based on a structure lifetime of 75 years. The required quality of concrete, as well as cover thickness of concrete, is calculated according to the different weather conditions in which the structure exists (Table 5.13).

5.6 DESIGN PRECAUTIONS FOR CHLORIDE EFFECTS

This section will clarify the impact of the design on the concrete mix and the spread of chlorides into the concrete with regard to weather factors. Chloride distribution will be studied to determine the lifetime of a structure under the influence of chlorides. Various atmospheric factors affect concrete structures. For example, although bridges and parking garages are exposed to air, concrete pilings are exposed to seawater. Therefore, the location of a structure is very important because of the weather factors affecting it.

We notice that the lower requirements of the various codes are not enough to preserve the life of a structure; however, it can be maintained impermeable by adding with some additives to protect against corrosion. The basis of the design depends on the chloride content at different times, and the expense during the lifetime can be estimated in a simple way by the relationship between the concentration of chlorides and depth at a specific time, as in Figure 5.4.

Three main parameters affect the calculations:

1. The effect of the diffusion factor, D_{eff}, depends mainly on concrete quality.
2. The effect of chloride concentration depends on the surface and CO_2 is mainly affected by the weather conditions surrounding the structure and can also be affected slightly by the cement content in concrete.
3. The chloride concentration that causes corrosion needs to be limited. There are two values involved in corrosion. The first value is required to start corrosion (C_i) and the second value is required to spall the concrete cover (C_p).

To provide good design and concrete quality, these three factors have been calculated by the approximations of Bamforth (1994) and Berke and Hicks (1992). This will be explained in the following sections.

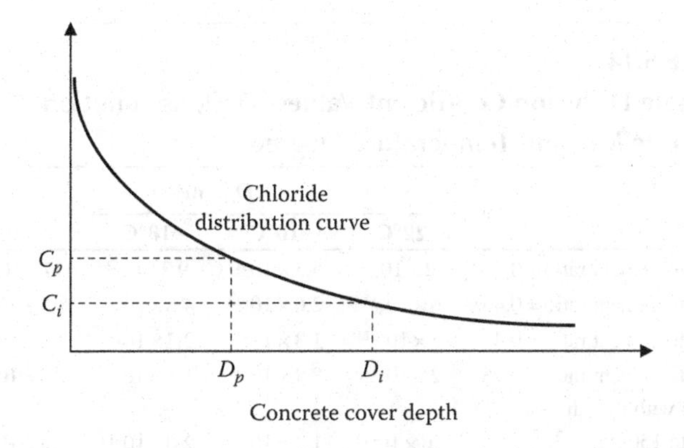

FIGURE 5.4 Sketch represents the relation between chloride concentration and depth at time, *T*.

5.6.1 Effect of the Diffusion Factor on Chloride Permeability

There are several methods to calculate the diffusion coefficient, depending on the type of concrete that is used or, in cases of an unknown type of concrete, nonavailability of data. In the absence of previous concrete data or a time by which to define the chloride content, it is possible to calculate the coefficient by measuring the resistance of concrete to electric conductivity. This is appropriate since the relationship between the resistance of electricity, written with the value of a coulomb (C) and increased chloride permeability test (ASTM C 1204094), works with the many tests. Through data analysis, the value of D_{eff} is obtained through the following two equations:

$$D_{eff} = 54.6 \times 10^{-8} (\text{Resistivity})^{-1.01} r^2 = 0.95$$

$$D_{eff} = 0.0103 \times 10^{-8} (\text{Rapid permeability,coulomb})^{-0.84} r^2 = 0.97$$

The resistivity is measured in units of kΩ·cm and permeability is measured in units of coulombs. However, D_{eff} is measured by units of square centimeters per second (cm^2/s). We can note from Table 6.1 that the temperature factor affects the spread of chlorides and therefore the temperature must be taken into account; this was not taken into account in the previous equations.

In the absence of any concrete data to be used, there is some general information that could be indicative of the value of D_{eff} through a combination of information available from Bamforth (1994) and Berke and Hicks (1992) because they relied on D_{eff}, which directly affected the quality of concrete. Bamforth provided a direct relationship between the diffusion factor and concrete compressive strength, as shown in Table 5.8, whereas Berke and Hicks clarified the relationship through the proportion of water to cement (Table 5.14). From Table 5.14, the values of the diffusion factor at different temperatures covering warm and cold weather are approximate.

TABLE 5.14
Estimate Diffusion Coefficient Values, Deff, as Function of Concrete Mix and Temperature Degree

Mix	D_{eff} (m²/s)			
	22°C	10°C	18°C	27°C
Water-to-cement ratio = 0.5	12×10^{-12}	5.3×10^{-12}	9.2×0^{-12}	17×10^{-12}
Water-to-cement ratio = 0.45	6×10^{-12}	2.6×10^{-12}	4.6×10^{-12}	8.3×10^{-12}
Water-to-cement ratio = 0.4	3×10^{-12}	1.3×10^{-12}	2.3×10^{-12}	4.2×10^{-12}
Water-to-cement ratio = 0.35 or 0.4 with fly ash	2×10^{-12}	8.8×10^{-13}	1.5×10^{-12}	2.8×10^{-12}
Equal to 1500 C	2.6×10^{-12}	1.2×10^{-12}	2.0×10^{-12}	3.4×10^{-12}
Equal to 1000 C	1.9×10^{-12}	8.3×10^{-13}	1.4×10^{-12}	2.6×10^{-12}
Equal to 600 C	1.3×10^{-12}	5.7×10^{-13}	9.9×10^{-13}	1.8×10^{-12}
Equal to 300 C	0.75×10^{-12}	3.3×10^{-13}	5.7×10^{-13}	1.04×10^{-12}

By studying Table 5.14, we will find that an increase of temperature increases permeability of chlorides. Therefore, in the case of designing a reinforced concrete structure in warm air, it is necessary to make the concrete with very low permeability. It has been found that in concrete cast from ordinary portland cement, only the permeability coefficient is affected significantly with structural age. However, in the case of concrete to which metallic additives such as silica fume and fly ash have been added, the coefficient of diffusion is influenced mainly by w/c ratio, as well as strength—especially under the influence of weather factors for a long period of time.

5.6.2 Effect of Chloride Concentration on Surface

Chloride concentration on the surface is directly affected by weather factors, and the value of chloride concentration on the surface is measured in units of weight and percentage or kilograms per cubic meter (kg/m³) of concrete. Berke and Hicks (1992) emphasize that these values function in the weather factors directly; various field studies have used the following values in design considerations:

- *Offshore structures*: Chlorides speed up to a fixed value of 18.7 kg/m³.
- *Structures prone to saline air*: Chlorides are an average increase of 0.15 kg/m³/year to the maximum value of 15 kg/m³/year.

Bamforth (1994) studied the impact of concrete components on chloride concentration on the surface. He found that with the increase in the cement content, the percentage of chlorides on the surface will increase. If the cement content increases from 200 to 400 kg/m³, the chloride content will be 0.5% to 0.6%, with about 20% increase. However, when the content of cement increases from 400 to 600 kg/m³, chloride content increases from 0.6% to 1.2% at 100%. Thus, chloride content on the surface must be taken into account in cases of cement content equal to or more than 400 kg/m³.

5.6.3 Effective Chloride Concentration

The concept of the initial concentration of chlorides affecting corrosion has been studied. The value of chloride concentration is necessary for the beginning of corrosion or for breaking a passive protection layer. In laboratory analysis and in the field, it has been proposed that the time needed for the spread of corrosion in steel until the concrete cover cracks and falls are based on specific content per unit of mass, C_p. C_i and C_p are units as a percentage of the weight of cement, or kg/m^3 of concrete. Bamforth (1994) proposed these values:

$$C_i = 0.4\% C_p = 1.5\% \text{of cement weight}$$

$$C_i = 0.9\text{kg/m}^3 C_p = 3\text{kg/m}^3$$

From a scientific point of view, these are approximately the same values. In the previous case of the concrete using a cement content of 300 kg/m^3, the suggested values by Bamforth are C_i = 1.2 kg/m^3 and C_p = 3 kg/m^3. These are considered the same values as those suggested by Berke and Hicks (1992).

5.6.4 Calculating Structure Lifetime

From the previous equations, one can calculate the structure lifetime or determine the required thickness of the concrete cover, as well as the quality of concrete which has a high life expectancy, on the basis of design. Bamforth used a model and laboratory data for the calculation of the life expectancy of concrete, which is exposed to atmospheric factors in Table 4.8, using the British code. The lifetime of a structure in the case (C_p = 1%) is often less than 20 years.

The required quality of concrete expressed by D_{eff} and the thickness of the concrete cover for a structure lifetime of 50 years (or 120 years in the case of exposure to

TABLE 5.15
Concrete Cover Thickness to Achieve 50- or 120-Year Lifetime for Building in a Chloride Environment

Maximum Diffusion Coefficient				Concrete Cover (mm)
C_p = 1%		C_i = 0.4%		
Age = 120 years	Age = 50 years	Age = 120 years	Age = 50 years	
9.06×10^{-14}	2.17×10^{-13}	4.42×10^{-14}	1.06×10^{-13}	30
1.61×10^{-13}	3.87×10^{-13}	7.86×10^{-14}	1.89×10^{-13}	40
2.52×10^{-13}	6.04×10^{-13}	1.23×10^{-13}	2.95×10^{-13}	50
4.25×10^{-13}	1.02×10^{-12}	2.07×10^{-13}	4.98×10^{-13}	65
5.66×10^{-13}	1.36×10^{-12}	2.76×10^{-13}	6.63×10^{-13}	75
8.16×10^{-13}	1.96×10^{-12}	3.98×10^{-13}	9.54×10^{-13}	90
1.01×10^{-12}	2.42×10^{-12}	4.91×10^{-13}	1.18×10^{-12}	100

TABLE 5.16
Concrete Mix Limit[a]

	No Risk from Corrosion	Corrosion due to Carbonation				Corrosion due to Seawater Chloride			Corrosion due to Chloride without Seawater			Corrosion due to Chemical Attack		
	XO	XC1	XC2	XC3	XC4	XS1	XS2	XS3	XD1	XD2	XD3	XA1	XA2	XA3
Maximum w/c ratio		0.65	0.6	0.55	0.5	0.5	0.45	0.45	0.55	0.55	0.45	0.55	0.5	0.45
Concrete compressive strength	C12/15	C20/25	C25/30	C30/37	C30/37	C30/37	C35/45	C35/45	C30/37	C35/45	C35/45	C30/37	C30/37	C35/45
Minimum cement content (kg/m^3)		260	280	280	300	300	320	340	300	300	320	300	320	360

[a] Based on ENV 206.

TABLE 5.17
Different Degrees of Effect for Different Environmental Conditions

Deterioration Type	Exposed Degree	Surrounding Environment	Example
No possibility for corrosion	XO	No effect on concrete	Concrete inside the building with very little relative humidity
Corrosion due to carbonation	XC1	Dry	Concrete inside the building with very little relative humidity
	XC2	Wet and rarely dry	Some foundations or parts from retaining wall to water
	XC3	Moderate relative humidity	Concrete inside the building with moderate relative humidity or outside protected concrete from rain
	XC4	Dry and wet cycles	Exposed to water, not like XC2
Corrosion due to chlorides in the sea	XS1	Air with salts but not on coast	Building near the coast
	XS2	Immersed	Part of offshore structure
	XS3	Splash zone	Part of offshore structures
Corrosion due to chlorides	XD1	Moderate relative humidity	Concrete surface exposed to water spray containing chloride
	XD2	Wet and rarely dry	Swimming pool or concrete exposed to water from industrial waste
	XD3	Dry and wet cycles	Parts of bridges
Freezing and melting	XF1	Moderate saturation without ice melting	Main surface exposed to rain
	XF2	Moderate saturation with ice melting	Main surface on the road
	XF3	High water saturation without melting ice	Horizontal surface exposed to ice
	XF4	High water saturation with melting ice	Roads, bridges, decks, or main surface exposed to melting ice
Chemical attack	XA1	Exposed to slight chloride attack	
	XA2	Exposed to moderate chloride attack	
	XA3	Exposed to high chloride attack	

chloride) as calculated by Bamforth are shown in Table 5.15. The calculation is based on the beginning of corrosion C_i at 4% of the weight of cement and cracked concrete at the concentration of $C_p = 1\%$ of the weight of the cement. To obtain a thickness of concrete cover of less than 50 mm, the D_{eff} must be less than 10^{-13}, which can be obtained for a high quality of concrete with additives.

The value of the concrete diffusion coefficient, D_{eff}, for concrete with w/c ratio is equal to 0.45%, which is required by the code. A structure affected by bad weather and the increasing concentration of chlorides did not satisfy the requirements of the code as the previous one with 50–75 mm thickness of the concrete cover.

Predicting the structure lifetime was addressed by El-Reedy (2012) by using the reliability analysis technique.

5.6.5 General Design Considerations

It is clear that when any structure is designed, it is necessary to identify accurately the concrete quality and the degree to which it is exposed to the weather factors affecting it over its lifetime; these cause steel corrosion and therefore determine the thickness of the concrete cover of each member on the structure. Note that the Egyptian code has carefully defined the various weather factors that affect concrete structures (also see Table 5.9). The code of the European Union has identified the weather conditions to which structures are exposed (Table 5.16). A set of specifications, European ENV 206 (1992), set final revisions in April 1997 (Bentur et al.) that are presented here only to clarify how best to design buildings to achieve concrete with a longer lifetime.

In determining the forms of structures, it is necessary to find out the structure of the concrete, according to given value characteristics of the concrete. The required percentage of water-to-cement ratio must not exceed it and the cement content shall not be less than as stated in Table 5.17. Therefore, generally, to achieve the longest possible lifetime of a building, one must determine the thickness of the concrete cover, concrete characteristic compressive strength, and cement content in the concrete, as well as control the crack width to be consistent with the code followed in the design.

REFERENCES

ACI 318R-89. 1994. Building code requirements for reinforced concrete (ACI 318–89) and commentary (ACI 318R-89). In *Manual of Concrete Practice, Part 3*. Detroit, MI: American Concrete Institute.

ACI Committee 201. 1994. Guide to durable concrete. In *Manual of Concrete Practice, Part 1*. Detroit, MI: American Concrete Institute.

ACI Committee 222. 1994. Corrosion of metals in concrete. In *Manual of Concrete Practice, Part 1*. Detroit, MI: American Concrete Institute.

ACI Committee 224. 1980. Control of cracking in concrete structures. *Concrete International—Design and Construction* 2(*10*):35–76.

ACI Committee 301. 1994. Specification for structural concrete for building. In *Manual of Concrete Practice, Part 3*. Detroit, MI: American Concrete Institute.

ACI Committee 357. 1994. *Guide for Design and Construction of Fixed Off-Shore Concrete.* Detroit, MI: American Concrete Institute.

Bamforth, P. B. 1994. Specification and design of concrete for the protection of reinforcement in chloride-contaminated environments. Paper presented at UK Corrosion & Eurocorr '94, Bournemouth International Center, Bournemouth, UK, October 31–November 3, 1994.

Beeby, A. W. 1979. *Concrete in the Oceans—Cracking and Corrosion. Technical Report No. 2 CIRIA/UEG.* Slough, UK: Cement and Concrete Association.

Bentur, A. and C. Jaegermann. 1989. *Effect of Curing in Hot Environments on the Properties of the Concrete Skin (in Hebrew). Research Report, Building Research Station.* Haifa, Israel: Technion—Israel Institute of Technology.

Bentur, A. and C. Jaegermann. 1990. Effect of curing and composition on the development of properties in the outer skin of concrete. *ASCE Journal of Materials in Civil Engineering 3(4)*:252–262.

Bentur, A., S. Diamond, and N. S. Berke 1997. *Steel Corrosion in Concrete.* London, UK: E & FN Spon.

Berke, N. S. and M. C. Hicks. 1992. The life cycle of reinforced concrete decks and marine piles using laboratory diffusion and corrosion data. In *Corrosion Forms and Control for Infrastructure*, ASTM STP 1137, pp. 207–231, ed. V. Chaker. Philadelphia, PA: American Society for Testing and Materials.

Browne, R. D., M. P. Geoghegan, and A. F. Baker. 1993. Analysis of structural condition from durability results. In *Corrosion of Reinforcement in Concrete Construction*, pp. 193–222, ed. A. P. Crane. London, UK: Society of Chemical Industry.

BS 8110. 1985. *Structural Use of Concrete, Part 1. Code of Practice for Design and Construction.* London, UK: British Standard Institution.

El-Reedy, M. A. 2012. *Reinforced Concrete Structure Reliability.* Boca Raton, FL: CRC Press.

European Prestandard ENV 206. 1992. *Concrete—Performance, Production, Placing and Compliance Criteria.* London, UK: BSI.

Parrot, P. J. 1994. Design for avoiding damage due to carbonation-induced corrosion. In *Durability of Concrete*, ACI SP-145, pp. 283–298, ed. V. M. Malhotra. Detroit, MI: American Concrete Institute.

Van Daveer, J. R. 1975. Techniques for evaluating reinforced concrete bridge decks. *Journal of the American Concrete Institute* 72(*12*):697–704.

6 Controlling Corrosion in Steel Bars

6.1 INTRODUCTION

Controlling the corrosion process in reinforcement steel bars buried in concrete is accomplished by controlling as best as possible all the factors that help to start the corrosion and spread it throughout the steel reinforcements in the concrete structure. All codes have different requirements and specifications concerning the required thickness of the concrete cover, the water-to-cement (w/c) ratio in the mixture, and cement content based on the weather factors surrounding the structure. These requirements have been included in the code after many field and laboratory tests have been performed in order to formulate specific guidelines and recommendations.

To protect concrete structures from corrosion, one must bear in mind some precautions to be taken for the use of concrete specifications commensurate with the weather factors surrounding the structure. These will be listed in this chapter, where we will discuss how to control the corrosion process. As for corrosion protection for the structure, several methods and new advanced techniques will be clarified in detail in Chapter 7. These protection methods and techniques do not dismiss the need to control the corrosion process by using good quality-control procedures in concrete construction.

Different codes and specifications provide us the requirements and limits to be followed in the design and implementation of concrete structures, such as the thickness of the concrete cover and the proportion of chlorides in the mix, as explained in Chapter 5. However, this does not obviate the use of some other precautions during these phases of concrete mixing. For example, some countries have weather conditions that result in special economic situations, necessitating the development of many precautions during construction. Therefore, this chapter will explain how to control the process of corrosion by controlling the process of building the concrete structure. It notes some simple ways of taking precautions during construction at a lower cost compared to other simple ways of protection, which will be explained in Chapter 7. The cost of these requirements is an especially low cost that is not comparable to the high cost of the repair processes in the future.

It is necessary to control corrosion by taking into account the requirements of the construction; a suitable design will consider a very good method of protection without adding any extra cost to the project. The control of the corrosion process depends mainly on controlling the two factors causing the corrosion: the carbonation transformation process and spread of chloride into concrete. The control of these two operations occurs in three main ways: Study of the impact of weather factors on the process of corrosion, the component of the concrete mix and its impact on the process of corrosion, and the curing process for concrete. Understanding these

DOI: 10.1201/9781003407058-6

processes and their effects on the process of corrosion from carbonation transformation or chloride propagation will provide the ability to choose the required method of curing and appropriate mixing ratios to control corrosion as much as possible.

6.2 CARBONATION PROCESS CONTROL

The process of carbonation was clarified in Chapter 4 and it can be defined simply as the propagation of carbon dioxide into the concrete until the CO_2 reaches the steel bars. This process then breaks the passive protection layer around the steel bars, consequently, starting the corrosion process. The carbonation process is highly influenced by the weather conditions.

6.2.1 Effect of Environmental Conditions

The carbonation transformation process is affected greatly by the amount of moisture in concrete. This is very clear in Figure 6.1a and b and based on a study performed by Wierig (1984). From these figures, it is found that when relative humidity is very high, the concrete pore voids will be full of water and it is difficult for CO_2 to spread inside the concrete. Conversely, in cases of very low relative humidity, there will be little moisture, so the carbonation chemical reaction will be difficult and the

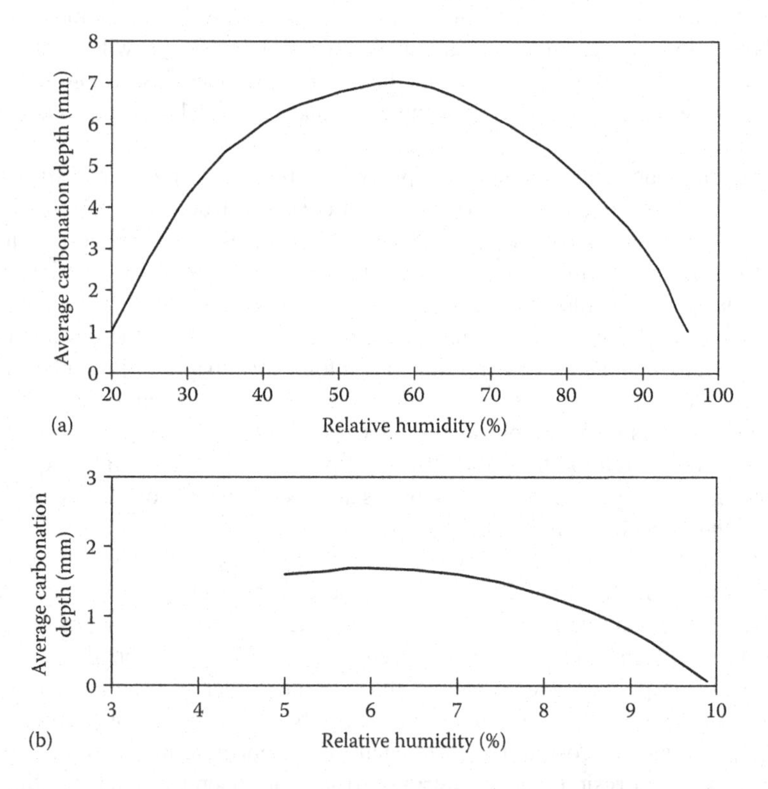

FIGURE 6.1 Effect of relative humidity to carbonation depth (a) at age 2 years with 0.6 w/c and (b) at age 16 years with w/c equal to 0.6.

carbonation transformation process will be slow. As shown in these two figures, the results are the same when the age of the building is 2 or 16 years.

We must differentiate between the relative humidity of the air space surrounding the structure and relative humidity inside the pore voids of concrete. Also, note that concrete tends to dry slowly with depth and that the moisture varies with the time inside the concrete cover. This shows that the calculation of the rate of carbonation transformation as a function of the relative humidity is not entirely accurate. Our inquiries can be answered, however, because many equations calculate the depth of carbonation and most of them are empirical equations; this was detailed in Table 3.1.

The depth of carbonation is also affected by other factors—for example, when concrete is exposed directly to the rain, or concrete in special conditions that protect the structure from exposure to rain, or the w/c ratio (Figure 6.2), according to studies performed by Tutti (1982). From Figure 6.2, one can find that the carbonation transformation depth increases with time. In any case, this depth is affected by w/c ratio. After 50 years, the carbonation transformation depth is equal to around 15 mm at w/c ratio equal to 0.45 and reaches over 50 mm when w/c ratio is equal to 0.75.

Concrete under a shelter that is protected from rain shows a greater carbonation depth than when it is without shelter and unprotected from rain. Calculating the rate of carbonation within the concrete through practical tests gives results that differ from those from application on-site; thus, in lab tests relative humidity is often between 40% and 60%, which is the region in which the carbonation process occurs quickly, as seen in Figure 6.1. Therefore, the laboratory tests give the worst cases of carbonation, as shown in Figure 6.3. From this figure, one can find that the relative carbonation depth increases in the laboratory more than in the field test. Also, protected concrete gives better results than those of unprotected concrete.

From this, one can conclude that the concrete inside a building, which is protected from the outside, is more highly influential in the carbonation process compared with exterior concrete exposed to the outside without protection—assuming that the same source of CO_2 is available in the two cases. However, it is usually the exterior member that is exposed to higher amounts of CO_2 due to the surrounding environmental conditions.

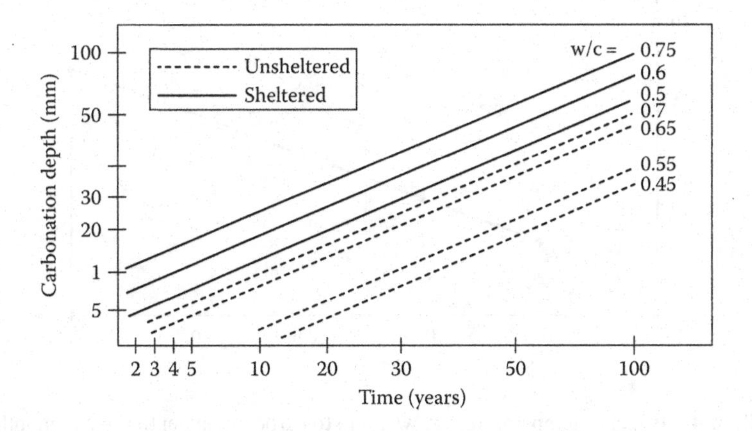

FIGURE 6.2 Relation between carbonation depth and w/c for different conditions.

One of the weather factors that affects the process of carbonation is the temperature. With increasing temperature, the rate of carbonation also increases, as shown in Figure 6.4. This figure shows that when the temperature is increased from 20°C to 30°C, the carbonation process is increased by 50%–100% with different w/c ratios. These tests are performed by placing the samples in the atmosphere for about 15 months. Note in this figure that a higher w/c ratio will result in increase in the carbonation depth. Also, for a given w/c ratio, the depth of carbonation is higher with the increase in temperature. This result is very important, especially in the Middle East, where countries are exposed to high temperatures with high chances of carbonation, especially in petrochemical plants. Therefore, as we will discuss maintenance strategy in Chapter 9, concrete structures that exist in higher ambient temperatures need less periodic maintenance time.

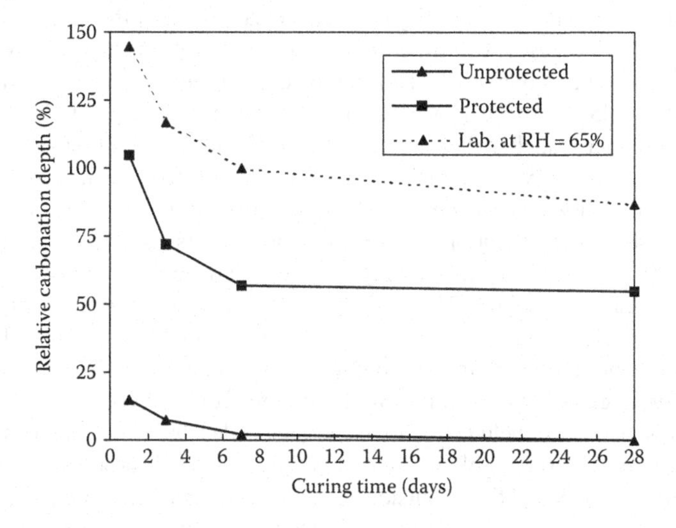

FIGURE 6.3 Curing effect and surrounding environment at carbonation depth of concrete aged 16 years.

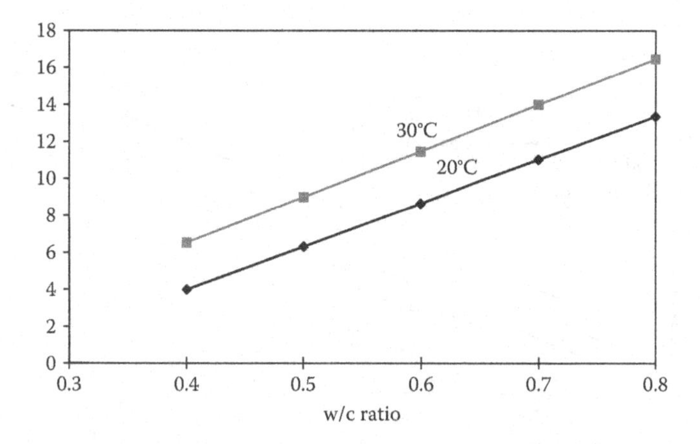

FIGURE 6.4 Effect of temperature and w/c ratio to carbonation depth after 15 months.

(From Jaegermann, C. and D. Carmel, Factors affecting the penetration of chlorides and depth of carbonation [in Hebrew with English summary], Final research report, Haifa, Israel: Building Research Station, Technion-Israel Institute of Technology, 1991.)

When developing the building assessment, it is important to know the temperature around the deteriorated concrete member as it can provide a reason for corrosion due to carbonation. Note that, even though when the structure is inspected the ambient temperature is good, the structure may have been exposed to high temperatures over a period of time. Therefore, it is important to collect the data about the ambient temperature affecting the concrete throughout its lifetime because this information could change the choice of repair and future maintenance plans for the building. Also, Figure 6.4 shows that at any ambient temperature, the carbonation transformation depth is increased by increasing the w/c ratio.

6.2.2 Components of Concrete Mix

Weather factors cannot be controlled by the engineer who does the construction at the site, but he can control the components of the concrete mixture as well as the cement content or the w/c ratio and the process of curing after casting. In general, the impact of the w/c ratio is very high, as shown in the previous figure. With the increase in the w/c ratio, the depth of carbonation increases. On the other hand, the impact of the use of slag furnaces (fly ash) is clear in Figure 6.5. As the figure shows, when fly ash is used, the depth of carbonation is higher than when ordinary portland cement is used. In both cases, increasing the w/c ratio will increase the depth of carbonation.

As stated in Hobbs's (1994) research study, with increasing concrete compressive strength, the carbonation depth will decrease. This test is performed at a building age of 8.3 years and is done for concrete containing ordinary portland cement; another test is done when fly ash is used. This relationship has been created through a curve that gives a rapid increase in the depth of the transformation of carbonation with a decrease in concrete compressive strength. Figure 6.6 shows the effect of the depth of carbon transformation by the concrete compressive strength for different

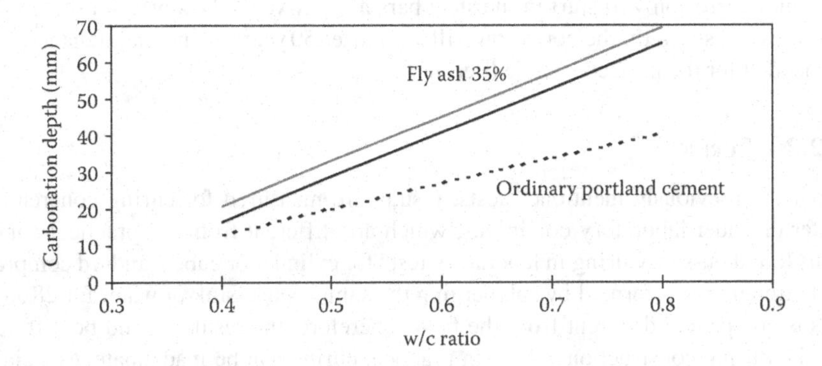

FIGURE 6.5 Relation between w/c ratio and carbonation depth after 100 months.

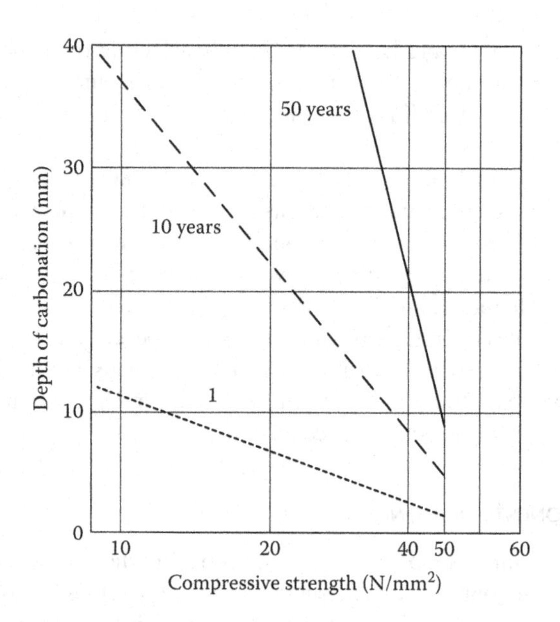

FIGURE 6.6 Effect of concrete compressive strength on carbonation depth at various ages.

ages of 1, 10, and 50 years. With the increase of the lifetime of a building, the depth of the carbonation transformation increases. This test was carried out by Smolczyk (1968). It can be noted that from the first year up to 10 years, the carbonation depth is decreased with increasing concrete compressive strength but does not decrease at the same rate at the age of 50 years.

It is important to know from Figure 6.6 that at 20 N/mm^2 carbonation transformation depth will be around 24 mm after 10 years, whereas in the same conditions at 40 N/mm^2, it will be 5 mm at 10 years; thus, the decrease in carbonation depth is about five times higher. Note in the figure that the carbonation depth is 20 mm at 50 years. This information is valuable when preparing a maintenance plan and selecting a suitable concrete compressive strength. Selecting a 20 N/mm^2 compressive strength is less expensive than 40 N/mm^2. On the other hand, for 20 N/mm^2 compressive strength, corrosion will start in the steel bars after 10 years; however, for 40 N/mm^2 compressive strength, the corrosion will start after 50 years, so no maintenance may be needed for the project's whole lifetime.

6.2.3 Curing

From the previously mentioned tests, results are measured for curing concrete by water or under laboratory conditions, which are different from the practical conditions found on-site. Curing in laboratory tests for cylinder or cube standard compressive strength is performed by submerging the samples in a sink of water for 28 days; this is completely different from the field. Therefore, the results would be different on-site during construction, when, in practice, curing can be inadequate, especially in areas with warm, dry air.

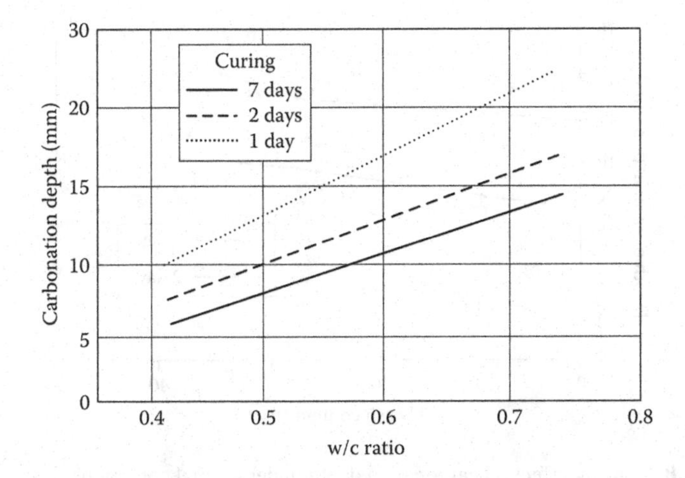

FIGURE 6.7 Effect of w/c ratio and curing to the carbonation depth.

The effect of curing duration and w/c ratio on the transformation of carbonation is also clarified in Figure 6.7. The figure reflects a coastal city in a warm, dry region in the Mediterranean Sea, where the temperature is 30°C and relative humidity is 40% after 5 years. Improper curing will increase the depth of the carbon transformation rate to two to four times that for a proper curing process. Note that the worst conditions do not assist in proper curing when the temperature is high for dry air and also for a moist atmosphere.

Therefore, we must care about concrete mixing ratio and follow up the quality of the execution; consequently, we must care about the w/c ratio and the curing process. The curing process is important for two reasons: Improper curing will decrease the concrete compressive strength by about 20% (Bentur and Jaegermann 1990), and it will have a significant impact on increasing the depth of carbonation transformation. The impact of improper curing of concrete is greater when fly ash is used in the concrete mix because the reaction is slow, as is obvious in Figure 6.8. With the increasing amount of fly ash, more depth of carbon transformation will be found; after 2 days, the rate of increasing carbonation transformation depth will be higher after increasing the percentage of fly ash content by about 30% (Figure 6.8).

The adverse effect of increasing the depth of the carbonation transformation with the increase of slag furnace content must be considered when buildings are exposed to the influence of carbonation. Therefore, the designer must consider not using fly ash in concrete if the building will be exposed to carbonation during its lifetime. Note that attention must be given to a curing process using fly ash and that the curing process varies when additives are used because their properties are different even when they are of the same nature (such as silica fumes, slag, and fly ash). These differences in their behavior do not exist in the specifications.

Therefore, in the case of mineral additives, the engineer must decide whether they will improve the quality of concrete and must also examine the appropriateness of the nature of the structure and of the work, as well as the weather conditions. It is already clear that inattention to the process of concrete curing after casting has a

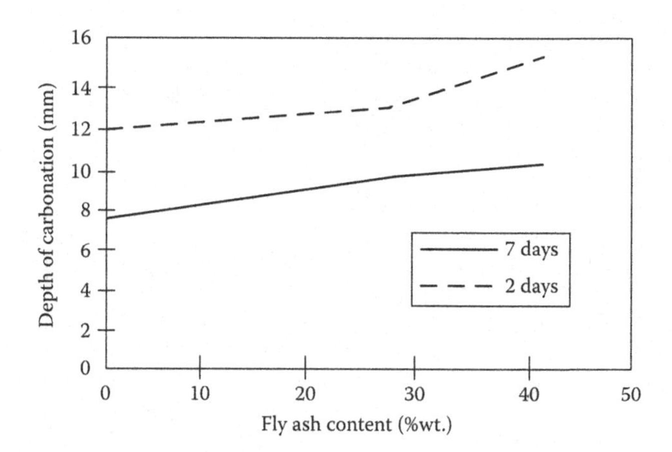

FIGURE 6.8 Curing effect on carbonation depth under normal conditions.

great impact on concrete compressive strength as well as on the carbonation process. From earlier figures, note that there is little difference in the effect of curing duration on the lack of depth of the carbon transformation. We know that the curing process certainly does not cost much when compared to the problems arising from the lack of a curing process.

In general, curing is important as it will enhance the reaction of the remaining cement particles that are not hydrated with the water during mixing. So, more reactions will increase the carbon oxides and potassium oxides, which increase the alkalinity of the concrete and prevent the propagation of chloride and carbonate.

6.3 CHLORIDE CONTROL

Chlorides are the second main cause of corrosion of steel bars embedded in concrete. The steel bars will be affected by the presence of chlorides inside the concrete itself or by chlorides from outside the structure, such as seawater, seawater spray, or a surrounding atmosphere saturated with salt. The different codes and specifications provide chloride limits in the concrete mix and confirm that the content of chlorides in the concrete has no effect on the start of corrosion. Moreover, this should reduce the chloride content over the lifetime of the structure.

6.3.1 Weather Factors Affecting Corrosion

Chlorides spread inside concrete as a result of chlorides inside the same concrete during casting or outside the structure due to seawater (e.g., an offshore structure). Chloride propagation inside concrete depends on the equation of Fick's second law, which was explained before, as well as on the diffusion factor. The natural factors that affect structures have a very significant impact on the probability of causing corrosion to the structure. This serious factor functions in the presence of chlorides in the surrounding environment, which could infiltrate the concrete. They must be paid proper attention to—for example, the splash zone in the case of offshore structures and other structures, such as bridges, where salt is used to melt ice.

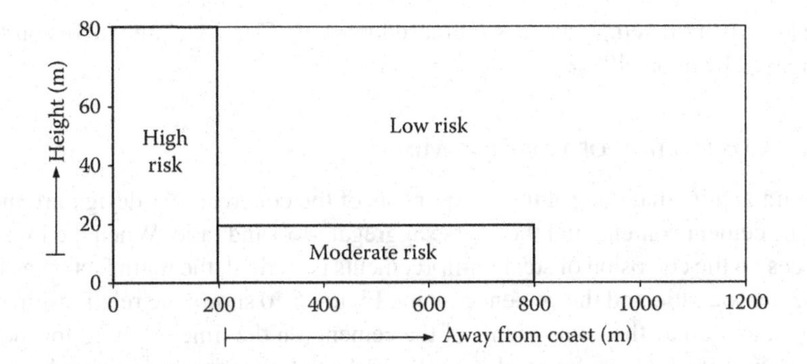

FIGURE 6.9 Structure corrosion risk based on the height and distance from coast.

Chloride can be in the form of a spray in the air when a marine environment surrounds the structure; an obvious example is that of a building under construction near the shore in a coastal city. Figure 6.9 shows the degree of serious threat posed to a structure according to its location in terms of its distance from the coast and also in terms of height. However, these data are not taken as a whole, but only as a guide to determine the degree of threat, depending on the climatic conditions of the region.

Another important factor is air temperature, especially in marine areas with warm atmospheres (e.g., the Arabian Gulf and the Middle East). Note the increase of the temperature diffusion coefficient in the following equation:

$$D_2 = D_1\left(\frac{T_1}{T_2}\right)e^{\left[k(1/T_1 - 1/T_2)\right]} \tag{6.1}$$

where D_2 and D_1 are the diffusion coefficients at temperatures T_2 and T_1, respectively. The results of laboratory tests clarified the relationship between the quality of concrete and increase of the temperature with the diffusion coefficient.

From this relationship and also as presented in Table 6.1, note that, with the increase of the temperature coefficient, the diffusion coefficient increases and that the diffusion factor increases about twofold in the event of an increase in temperature of about 10°C. At any temperature, the lower the w/c ratio is, the lower the coefficient of diffusion will be. Also, when silica fume is added to the concrete mix, it will reduce the diffusion factor by about half, as shown in the previous table.

TABLE 6.1
Effect of Weather Temperature and Concrete Type on the Diffusion Factor

	Diffusion Factor Effect ($\times 10^{-12}$ m^2/s)	
Concrete Type	**10°C**	**22°C**
Water-to-cement ratio = 0.4	1.3	3
Water-to-cement ratio = 0.3	0.9	2
Water-to-cement ratio = 0.4 and 7.5% Silica fumes	0.6	1.3

However, when the temperature is increased from 10°C to 20°C, the diffusion factor is increased by about 100%.

6.3.2 Composition of Concrete Mix

The main factors that control the components of the concrete mix design are the w/c ratio, the cement content, and the coarse aggregate-to-sand ratio. When the impact of chlorides on the corrosion of steel reinforcements is studied, the main factors affected will be the w/c ratio and the cement content. Figure 6.10 shows the relationship of the w/c ratio, as well as the components of the cement, on the time required for the start of corrosion. Rasheduzzafar et al. (1990) conducted experiments in which samples of concrete were partly submerged in a 5% solution of sodium chloride. From this figure, one can find that, with the increase in w/c ratio, the time to the beginning of corrosion will be decreased. In addition, the use of sulfate-resistant cement reduces the time required for corrosion to begin.

Verbeck (1975) stated that an increase of the C_3A content in portland cement will delay the deterioration and fall of concrete due to cracks resulting from corrosion in the steel reinforcement for the structures exposed to seawater. The different codes give the minimum limit of cement content required, but it is obvious that with the increase in cement content, the chloride diffusion factor is decreased at the same w/c ratio, as shown in Figure 6.11. When the w/c ratio is equal to 0.4, the effect of increasing the cement content in the concrete mix on the diffusion factor will decrease. Therefore, the cement content and the w/c ratio are the main factors that control the degree of chloride propagation inside the concrete. Figure 6.11 presents the test results of samples exposed to seawater according to research by Pollock (1985).

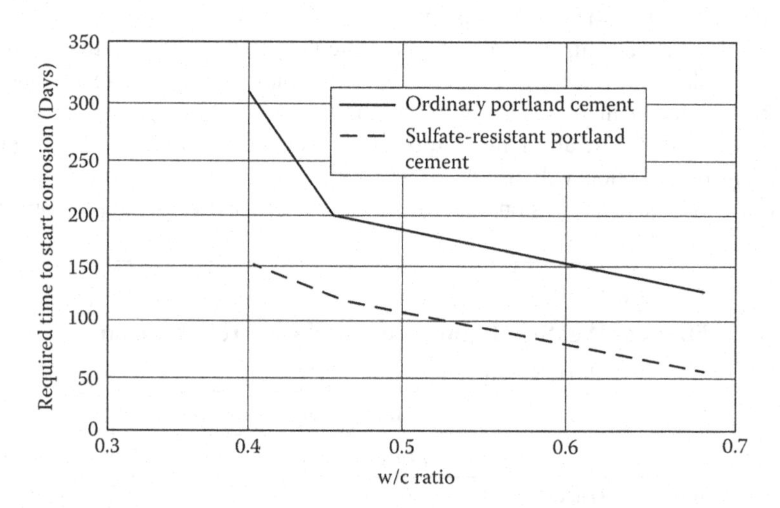

FIGURE 6.10 Relation between start corrosion and w/c ratio for different types of cement.

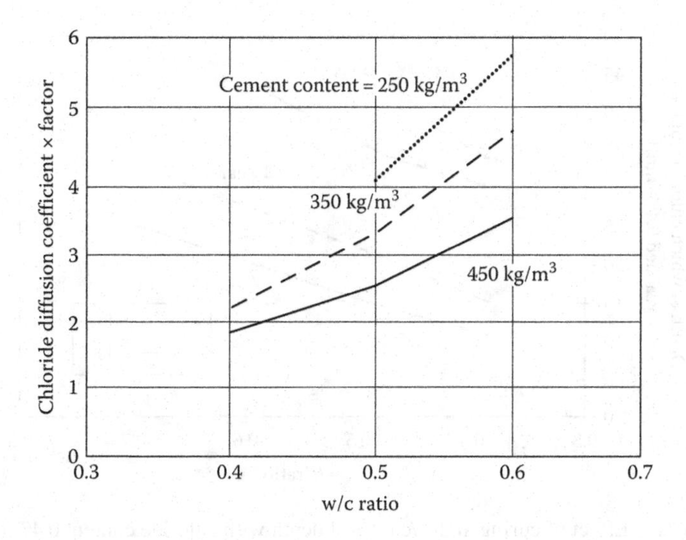

FIGURE 6.11 Effect of w/c and cement content to chloride propagation factor in concrete exposed to sea water.

6.3.3 Curing for Chloride Attack

As we stated in the case of the carbonation process, as well as in the case of chlorides, proper curing can have a significant impact, as it affects the spread of chlorides in the concrete, especially in the case of structures exposed to marine environments and warm air. The relationship of the depth of concrete containing 0.4% chloride in the range of 1–5 years and a w/c ratio equal to 0.5 is presented in Figure 6.12 (Jaegermann and Carmel 1988, 1991). This figure shows that, in any case, the depth of chloride in a curing process performed in 7 days is less than that cured for only 2 days; hence, the curing process has a direct impact on the depth of chloride penetration in the concrete. Note also that during the first year, the impact of curing on the depth of propagation of chlorides is higher than that in effect after about 5 years.

When comparing carbon dioxide propagation within concrete or chloride propagation and rear access to depth at 0.4% concentration of chlorides in the same period of curing and with the same w/c ratio, note that the chloride spreads faster than carbon dioxide does. Comparing the effects of using ordinary portland cement or cement containing fly ash with 25% by weight of cement, based on Jaegermann and Carmel (1988), shows that the case of quick curing and the w/c equal to 0.5 after 1 year of exposure to warm sea weather. We can note an increasing percentage of the chloride content with depth within the concrete when fly ash is used compared to when ordinary portland cement is used. Therefore, when fly ash additives are used, the engineer must be sure to make the initial tests to determine the required quantity necessary not to have an impact on corrosion control on the steel reinforcement.

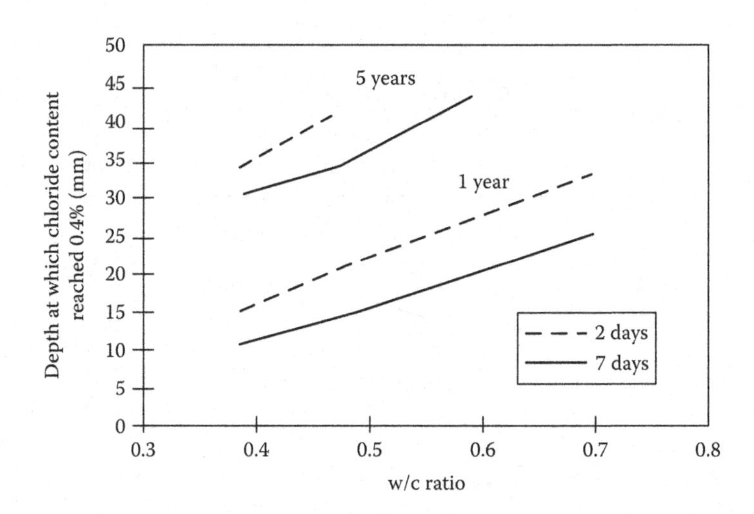

FIGURE 6.12 Effect of curing and w/c ratio at depth with chloride content 0.4% by cement weight in warm weather.

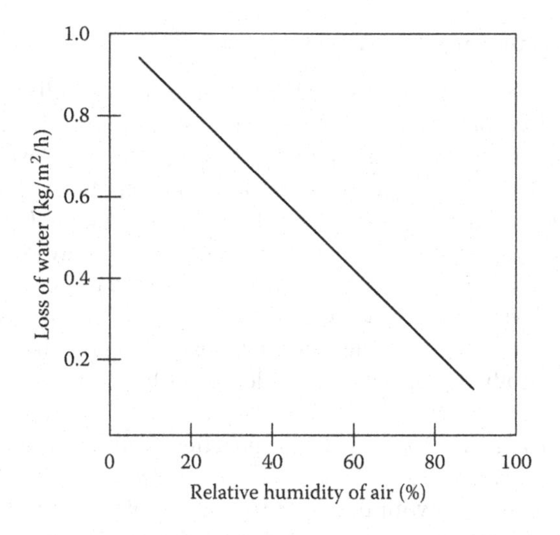

FIGURE 6.13 Relation between RH and loss of water. From Neville, A. M. 1981. *Properties of Concrete*. London, UK: Pittman Books Limited.

6.3.4 Execution of Curing

Recently poured concrete must be protected from rain and fast drying as a result of severe storms or warm, dry air. This is done by covering the cast concrete with a suitable coating system after finishing the pouring process until the time to final hardening. The curing process then proceeds in a manner already defined by the designer. Figures 6.13–6.15 illustrate the impact of relative humidity, air temperature, and wind speed on the speed of concrete drying.

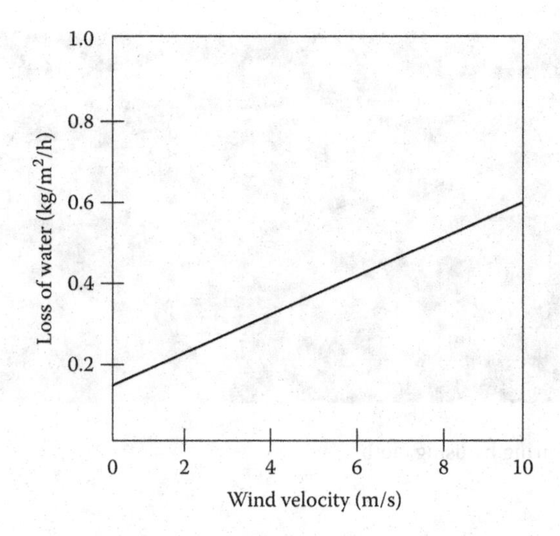

FIGURE 6.14 Relation between wind speed and loss of water. From Neville, A. M. 1981. *Properties of Concrete*. London, UK: Pittman Books Limited.

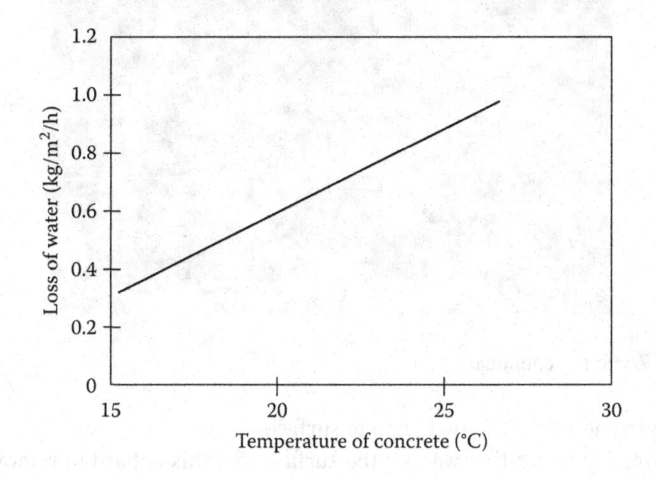

FIGURE 6.15 Relation between temperature and loss in water.

The purpose of the curing process is for the concrete to be wet for a period of not less than 7 days for ordinary portland cement and not less than 4 days in the case of fast-hardening cement or if additives have been used to accelerate the setting time. The curing process can be performed in many ways, including

- Spraying water that is free of salt and any harmful substances.
- Covering the concrete surface by a rough cloth, sand (around edges), or manufacturer wood waste and keeping it wet by spraying water on it regularly.
- Covering the concrete surface by high-density polyethylene sheets.

FIGURE 6.16 Curing by using cloth.

FIGURE 6.17 Spray chemicals.

- Spraying additives on the concrete surface.
- Painting chemical-like wax on the surface, but this is hard to remove, so it is preferred to be used in the foundation.
- Using steam for curing in some special structures.

Figures 6.16–6.18 represent different types of curing. Figure 6.16 illustrates the second common method of curing by covering the slab with a cloth that is always kept wet. The first curing method is spraying water on the concrete member in the early morning and at night, avoiding spraying water after sunrise as it will produce some cracks on a concrete surface due to evaporation of water. This method is used for normal environmental conditions. However, in some areas, such as the Middle East, the temperature in summer may reach as high as 55°C. In these conditions, different techniques are used, such as using some chemical to spray on the surface to prevent

FIGURE 6.18 Plastic sheet to protect from evaporation.

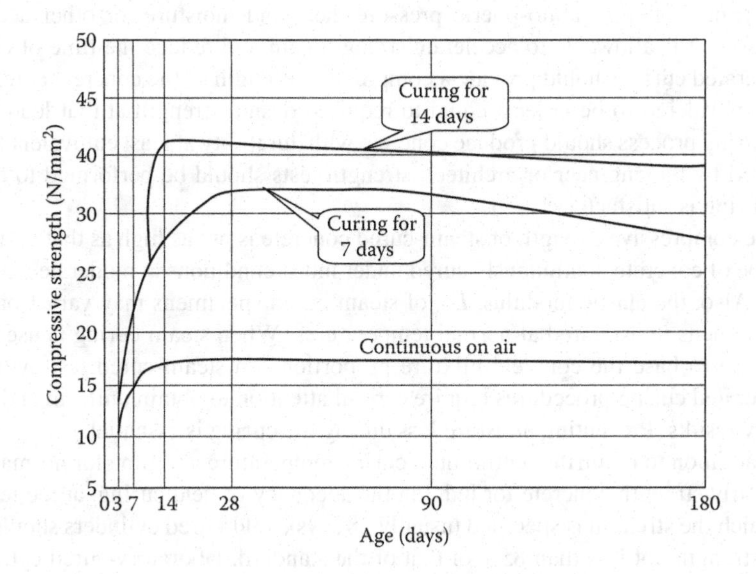

FIGURE 6.19 Sketch presents the relation between curing time and compressive strength.

evaporation of the water in the concrete mix, as shown in Figure 6.17. Figure 6.18 shows the other method: Covering the concrete member with a special plastic sheet to avoid the evaporation of water.

Another type of method is to spray the remnants of the wood manufacturing process, distribute them on the slab, and keep them wet all the time. A similar idea is to distribute sand on the slab and spray it with water so that it is always wet, taking care that the wet sand has a high dead weight that can affect the building from its recent construction. The curing process involves special issues: It costs little compared to the cost of concrete as a whole; it increases the resistance of concrete in a very significant way, that is, when care is taken during the curing process, the strength of the concrete will increase. Figure 6.19 shows that the concrete strength clearly increases when the curing process takes 7 days—more than that in the case of 3 days' curing. A stronger effect will be found when the duration of curing is 14 days, and this increase in concrete strength will be continuous throughout the structure's lifetime.

Moreover, the increase in concrete strength due to curing for 28 days is not more than that from the process of curing for only 14 days. The specifications of the project should provide clearly for a method of curing and define the required curing time exactly as they differ from project to project, according to the weather factors of the area in which the project is located and according to the concrete members.

6.3.4.1 Curing Process in ACI

According to ACI clause 5.11, the curing of concrete should be maintained above 50°F and under moist conditions for at least the first 7 days after placement; high early-strength concrete should be maintained above 50°F and under moist conditions for at least the first 3 days. If it is necessary to accelerate curing by high-pressure steam, steam at atmospheric pressure, heat and moisture, or other accepted processes, it is allowable to accelerate strength gain and reduce the time of curing. Accelerated curing should provide a compressive strength of the concrete at the load stage considered to be at least equal to required design strength at that load stage. The curing process should produce concrete with durability at least equivalent to that required by the engineer or architect; strength tests should be performed to ensure that curing is satisfactory.

The compressive strength of steam-cured concrete is not as high as that of a similar type of concrete continuously cured under moist conditions at moderate temperatures. Also, the elastic modulus, Ec, of steam-cured specimens may vary from that of specimens moist cured at normal temperatures. When steam curing is used, it is advisable to base the concrete mixture proportions on steam-cured test cylinders. Accelerated curing procedures require careful attention to obtain uniform and satisfactory results. Preventing moisture loss during the curing is essential.

In addition to requiring a minimum curing temperature and time for normal- and high early-strength concrete for judging the adequacy of field curing, at the test age for which the strength is specified (usually 28 days), field-cured cylinders should produce strength not less than 85% of that of the standard, laboratory-cured cylinders. For a reasonably valid comparison to be made, field-cured cylinders and companion laboratory-cured cylinders should come from the same sample. Field-cured cylinders should be cured under conditions identical to those of the structure—for example, if the structure is protected from the elements, the cylinder should be protected.

6.3.4.2 British Standard for Curing

Curing is the process of preventing the loss of moisture from concrete while maintaining a satisfactory temperature. The curing regimen should prevent the development of high-temperature gradients within the concrete. The rate of development of strength at early ages of concrete made with supersulfated cement is significantly reduced at lower temperatures. Supersulfated cement concrete is seriously affected by inadequate curing and the surface has to be kept moist for at least 4 days. Curing and protection should start immediately after the compaction of the concrete to protect it from the following:

- Premature drying, particularly by solar radiation and wind.
- Leaching out by rain and flowing water.
- Rapid cooling during the first few days after placement.
- High internal thermal gradients.
- Low temperature or frost.
- Vibration and impact, which may disrupt the concrete and interfere with its bonding to the reinforcement.

When structures are of considerable bulk or length, the cement content of the concrete is high, or the surface finish is critical, special or accelerated curing methods are to be applied. The method of curing should be specified in detail.

BS 1881 states that surfaces should normally be cured for a period of not less than that given in Table 6.2. Depending on the type of cement, the ambient conditions, and the temperature of the concrete, the appropriate period is taken from Table 6.2 or calculated from the last column of that table. During this period, at no part of the surface should the temperature fall below 5°C. The surface temperature is lowest as it rises and depends upon several factors, including the size and shape of the section, the cement class and cement content of the concrete, the insulation provided by the formwork or other covering, the temperature of the concrete at the time of placing, and the temperature and movement of the surrounding air. If not measured or calculated, the surface temperature should be assumed equal to the temperature of the surrounding air (see CIRIA Report 43 (1985)).

The most common methods of curing according to British specifications include

- Maintaining formwork in place.
- Covering the surface with an impermeable material such as polyethylene, which should be well sealed and fastened.
- Spraying the surface with an efficient curing membrane.

TABLE 6.2
Minimum Periods of Curing and Protection as in BS8110

Minimum Period of Curing (Days)			
Any Temperature between 10°C and 25°C	5°C–10°C	Condition after Casting	Type of Cement
$60/(t+10)$	4	Average	Portland cement and sulfate-resisting portland cement
$80/(t+10)$	6	Poor	
$80/(t+10)$	6	Average	All cement except the preceding and supersulfated cement
$140/(t+10)$	10	Poor	
No special requirements		Good	All

Notes: Good, damp, and protected (relative humidity greater than 80%; protected from sun and wind); average, intermediately between good and poor; poor, dry, or unprotected (relative humidity less than 50%; not protected from sun and wind).

6.4 PROTECTING SPECIAL STRUCTURES

The code and specification requirements in the design and execution of concrete structures to preserve the structure from corrosion are not sufficient in some structures exposed to very severe atmospheric conditions that cause corrosion, such as in offshore structures; special parts of the structure exposed to moving water (splash zone); or in parking garages or bridges, where salt is used to melt ice. There are some investments to establish modern commercial or residential buildings adjacent to seawater that might not be exposed to seawater directly; those buildings are exposed to chlorides, as well as some marine areas with warm atmospheres—for example, in the Middle East, especially the Arabian Gulf area.

Large investments have been made in structures on islands or quasi-islands with high temperatures, as well as in cities, such as Singapore and Hong Kong. Corrosion has already started and the corrosion rate will be very high. Therefore, control of corrosion through the use of good concrete and by determining the exact content of the cement, as well as the specific w/c ratio, and caution in the process of curing and during construction are key factors for protecting reinforced concrete structures from corrosion. These form the first line of defense in attacking corrosion. Nevertheless, in some structures, these methods are considered inadequate, so other methods are used to protect structures from concrete corrosion; these will be clarified in the next chapter.

However in general a good quality control is very important as per the case study for port Kembla saltwater swimming pool as per Sirivivatnanon (2019) as in his paper study the durability performance of Port Kembla Olympic Pool, built in 1937, has been investigated. Nearly all structural components were reinforced concrete and were exposed to marine environments with some components "permanently submerged" while others were in an "atmospheric zone" and "tidal or splash zone." After more than 60 years in service, most structural components were found to be in excellent condition. This paper discusses the site investigation that examined strength, carbonation, chloride penetration, and cover depths. The results revealed the quality of the concrete to be uniform in the pool but variable in other structural members. There was little carbonation but extensive chloride penetration, depending on the exposure condition. The average compressive strength of the 60-year-old concrete in the pool and its surrounding structures was 5700 and 4280 psi (40 and 30 MPa), respectively. The covers were between 2.0 and 2.5 in. (50 and 65 mm). Despite the extent of chloride penetration into the cover concrete, limited corrosion was observed. The concrete has proven to give a service life of over 60 years, which confirms the importance of achieving adequate strength and, perhaps more importantly, cover.

REFERENCES

Bentur, A. and C. Jaegermann. 1990. Effect of curing and composition on the development of properties in the outer skin of concrete. *ASCE Journal of Materials in Civil Engineering* *3*(4):252–262.

CIRIA Report 43. 1985. *Concrete Pressure on Formwork*. London, UK: Construction Industry Research and Information Association.

Hobbs, D. W. 1994. Carbonation of concrete containing PFA. *Magazine of Concrete Research* *46*(166):35–38.

Jaegermann, C. and D. Carmel. 1988. *Factors Affecting the Penetration of Chlorides and Depth of Carbonation (in Hebrew with English Summary). Research Report 1984–1987.* Haifa, Israel: Building Research Station, Technion-Israel Institute of Technology.

Jaegermann, C. and D. Carmel. 1991. *Factors Affecting the Penetration of Chlorides and Depth of Carbonation (in Hebrew with English Summary). Final Research Report.* Haifa, Israel: Building Research Station, Technion-Israel Institute of Technology.

Neville, A. M. 1981. *Properties of Concrete.* London, UK: Pittman Books Limited.

Pollock, D. J. 1985. Concrete durability tests using the gulf environment. In *Proceedings of the First International Conference on Deterioration and Repair of Reinforced Concrete in the Arabian Gulf,* The Bahrain Society of Engineers, Manama, Bahrain, Vol. I, pp. 427–441.

Rasheduzzafar, F. D., S. S. Al-Saadoun, A. S. Al-Gathani, and F. H. Dokhil. 1990. Effect of tricalcium aluminate content on corrosion of reinforced steel in concrete. *Cement and Concrete Research 20*(5):723–738.

Smolczyk, H. G. 1968. Carbonation of concrete—Written discussion. In *Proceedings of the Fifth International Symposium on Chemistry of Cement,* Tokyo, Japan, Part III, pp. 369–380.

Sirivivatnanon, V. 2019. Sixty-year service life of port Kembla saltwater concrete swimming pool. *ACI Materials Journal 116*(5):31–36.

Tutti, K. 1982. *Corrosion of Steel in Concrete.* Stockholm, Sweden: Swedish Cement and Concrete Research Institute.

Verbeck, G. J. 1975. Mechanisms of corrosion of steel in concrete. In *Corrosion of Metals in Concrete,* pp. 21–38. Detroit, MI: ACI Publication SP-49.

Wierig, H. J. 1984. Longtime studies on the carbonation of concrete under normal outdoor exposure. In *Proceedings of the RILEM Seminar on Durability of Concrete Structures under Normal Exposure,* Universitat Hannover, Hanover, Germany, pp. 239–253.

7 Methods for Protecting Steel Reinforcements

7.1 INTRODUCTION

The minimum requirements in various codes are often insufficient to ensure long-term durability of reinforced concrete exposed to severe weather conditions, such as those found in marine splash zones, bridges, and parking structures where deicing salts are applied. In addition, some newer structures (such as commercial buildings and condominiums) built-in marine areas, but not in splash zones, experience corrosion problems due to airborne chlorides. Furthermore, marine structures in the warmer climates prevalent in the Middle East, Singapore, Hong Kong, South Florida, etc., are especially vulnerable due to the high temperatures, which increase not only the rate of chloride ingress but also the corrosion rate once the process is initiated. In this chapter, a brief description of supplemental corrosion protection measures is given for structures, especially at risk. The use of good-quality concrete as described in the next chapters is considered the primary protection method, but various combinations of this with added supplements are necessary to reach the desired design life of the structure.

One of the most effective means to increase corrosion protection is to extend the time until a chloride or a carbonation front reaches the steel reinforcement. The minimum code requirements, which allow the use of concrete with a water-to-cement (w/c) ratio of less than 0.45 and concrete cover thickness of more than 38 mm, are totally inadequate for the structures and environmental conditions outlined previously if a design life of 40 or more years is specified. In many applications, designs complying with the minimum code requirements would not provide even as few as 10 years of repair-free service.

Taking precaution is better than repair: Protecting the concrete structure is easier and less expensive than repairing it. Repairing and renovating some parts of reinforced concrete buildings (e.g., foundations) is very expensive. In reality, protecting the structure from corrosion is protecting the investment during the structure's lifetime. In the previous chapter, the process of corrosion of reinforced steel bars in concrete was discussed for different environmental conditions; in addition, the different types of corrosion and their effects on the steel reinforcement were explained. In this chapter, our aim is to explain the different international methods available for protecting the reinforced concrete structure from corrosion.

Recently, more research and development have been conducted to provide economical and effective methods to protect steel reinforcement because this subject is very important for a construction investment considered to be worth a billion dollars worldwide. The preliminary method of protection of steel reinforcement is the achievement of good quality control in design and construction, taking all

DOI: 10.1201/9781003407058-7

precautions to avoid corrosion, as stated in the codes and specifications of different countries. These specifications vary with the different weather conditions to which a structure is exposed, as well as the function of the building. Offshore, underground, and surface structures are addressed by various important codes and specifications to protect the steel in their concrete members (see Chapter 5).

The second line of defense in protecting the steel bars is by using external methods. These methods include the use of

- Galvanized steel bars, epoxy-coated steel bars, or stainless steel.
- Additives, such as those used in cathodic protection, which are added to concrete during the pouring process.
- External membranes to prevent water permeability.
- Cathodic protection.

These methods have both advantages and disadvantages. This chapter discusses them and provides ways of application of each type of protection.

7.2 CORROSION INHIBITORS

There are two types of corrosion inhibitors: anodic and cathodic. The anodic inhibitor depends on the protection using a passive protection layer on steel reinforcement. Cathodic protection is based on preventing the spreading of oxygen in the concrete. Anodic inhibitors provide more effective protection than cathodic inhibitors, and they are commonly used in practice. Next, we will explain each type of protection, along with its advantages and disadvantages.

7.2.1 Anodic Inhibitors

The most common anodic inhibitor is calcium nitrate, which is well known for its compatibility with the process of pouring concrete at the site and has no adverse impact on the properties of concrete, whether it is fresh or in a hardened state. Other inhibitors include sodium nitrate and potassium nitrate, which are highly efficient in the prevention of corrosion; however, they are not used because, in cases of existing aggregates with alkaline, they react with cement, causing many problems and extensive damage to the concrete. Broadly, calcium nitrate has been used widely since the mid-1970s (Bentur et al. 1998). Note that calcium nitrate accelerates the concrete setting time. Broomfield (1995) mentioned that some more retardants must be added to the concrete mix at the mixing plant.

The mechanism by which calcium nitrate inhibits corrosion is associated with the stabilization of the passivation film, which tends to be disrupted when chloride ions are present at the steel level. The destabilization of the passivation film by chlorides is largely due to interference with the process of converting the ferrous oxide to the more stable ferric oxide. The anodic materials are used with the concrete exposed to chlorides directly, as in contact with seawater. The corrosion inhibitor reacts with chlorides and therefore increases the chloride concentration necessary for corrosion, based on the tests described in Table 7.1.

TABLE 7.1
Amount of Calcium Nitrate Required to Protect Steel Reinforcements from Chloride Corrosion

Calcium Nitrate Amount (kg/m^3 and 30% Solution)	Amount of Chloride Ions on the Steel (kg/m^3)
10	3.6
15	5.9
20	7.7
25	8.9
30	9.9

The amount of calcium nitrate to be added is determined based on the amount of chloride exposed to concrete; this can be done with practice or through knowledge of the quantity of chloride from previous experience. The addition of corrosion inhibitors does not diminish the importance of quality control of concrete and maintains the appropriate thickness of the concrete cover. In the case of high-strength, high-quality concrete, choosing and constructing the right concrete cover thickness and appropriate density of concrete according to specifications may obviate the need for a corrosion inhibitor for 20 years although a corrosion inhibitor is used when structures are exposed to chlorides directly (e.g., in offshore structures).

7.2.2 Cathodic Inhibitors

Cathodic inhibitors are added to the concrete during mixing. A new type of cathodic inhibitor coating is painted on the concrete surface after it hardens. This propagates to the steel bars through the concrete porosity and provides isolation, reducing the quantity of oxygen, an important driver in the corrosion process that spreads to concrete.

Many tests are performed for corrosion inhibition based on ASTM G109-92; their purpose is to define the effect of chemical additives on the corrosion of steel reinforcements embedded in concrete. From these tests, one finds that the corrosion inhibitor enhances the protection of the steel bars as well as controls concrete quality. However, cathodic inhibitors have less of an effect than anodic inhibitors. These tests have shown that, to obtain a higher efficiency in cathodic protection for the reinforced steel bars, it is essential to add a very large quantity of cathodic inhibitor to the concrete mix. However, cathodic inhibitors, such as amines, phosphates, and zinc, slow down the setting time considerably, especially when large quantities of cathodic inhibitors are used. Thus, when one decides to use a cathodic inhibitor, the concrete setting time must be taken into consideration.

From the previous discussion, one can conclude that an anodic inhibitor is more effective than a cathodic inhibitor. Therefore, in practice, the anodic inhibitor is generally used. If a cathodic inhibitor is used, the increase in concrete setting time should be kept in mind.

There are many cathodic inhibitors on the market with different commercial names and there was a study done by Newston and Robertsson (2019). This study compared different types and this study was a 10-year laboratory and field monitoring study to evaluate commercially available corrosion-inhibiting admixtures in terms of effectiveness at preventing or delaying corrosion. This study was using Xypex Admix C-2000, latex modifier, Kryton KIM, fly ash, silica fume, Darex Corrosion Inhibitor (DCI), Rheocrete CNI, Rheocrete 222+, and FerroGard 901. The laboratory study used accelerated corrosion testing (ASTM G109) to study 656 specimens from 100 different mixtures. Saltwater ponding and drying cycles continued until macrocell current readings indicated initiation of corrosion. Specimens were then removed from cycling and autopsied to determine the extent and location of corrosion on the reinforcing steel. After 10 years of cycling, all remaining specimens were autopsied. In addition to that, in their research, there was a field study that included 25 reinforced concrete panels exposed to the tidal zone in the Honolulu Harbor. Panels were tested regularly for chloride penetration and half-cell corrosion potential. After 10 years, the panels were removed from the site and examined. This study revealed that DCI, CNI, fly ash, and silica fume improved corrosion protection compared to control specimens. Concrete mixtures using Rheocrete 222+, FerroGard 901, Xypex Admix C-2000, or latex modifier showed mixed results.

7.3 EPOXY COATING OF STEEL REINFORCEMENT

It is important to paint steel bars using certain types of epoxies able to protect steel from corrosion. This method has yielded positive results, especially in steel exposed to seawater, in a study performed by the Federal Highways Association (FHWA) that has been evaluating the use of epoxy to coat steel reinforcements exposed to chloride attack. Also, some research studies, such as that conducted by Pike et al. (1972), Cairns (1992), and Satake et al. (1983), have demonstrated the importance of painting steel reinforcements. Epoxies have been used in painting reinforced steel for bridges and offshore structures since 1970.

Some shortcomings have been found in using this method. Precautions must be taken during the manufacturing and painting of steel, such as avoiding the absence of any friction between the bars, which would result in the erosion of the coating layer due to friction. Also, it is difficult to use methods for measuring the corrosion rate, such as polarization or half-cell potential, so it is not easy to predict the steel corrosion performance or measure the corrosion rate.

Painting steel-reinforced bars has been used extensively in the United States and Canada for 25 years. More than 100,000 buildings use coated bars, which is equal to 2 million tons of epoxy-coated bars. The coated steel bar must follow ASTM A 775M/77M-93, which sets allowable limits as the following:

- Coating thickness should be in the range of 130–300 μm.
- Bending of the coated bar around a standard mandrel should not lead to formation of cracks in the epoxy coating.
- The number of pinhole defects should not be more than six per meter.
- The damaged area on the bar should not exceed 2%.

These deficiencies cited by the code are the result of operation, transportation, and storage. There are some precautions that must be taken in these phases to avoid cracks in the paint. Andrade et al. (1994), Gustafson and Neff (1994), and Cairns (1992) define ways of storing and steel reinforcement bending, carrying the steel, and pouring concrete.

Painting steel reinforcement bars will reduce the bond between the concrete and steel; therefore, it is necessary to increase the development length of the steel bars to overcome this reduction in bond strength. According to American Concrete Institute (ACI) code (ACI Committee 318 1988), the increase of the development length is from about 20%–50%. The American code stipulates that, in the case of painting, the development length of steel bars must be increased by 50% when the concrete cover is less than three times the steel bar diameter or the distance between the steel bars is less than six times the bar diameter; in other circumstances, the development length should be increased by 20%.

Egyptian code does not take painting of steel into account. A study by El-Reedy, Sirag, and El-Hakim (1995) found that the calculation of the development length in the Egyptian code can be applied in the case of painting steel bars by epoxies without increasing the development length. However, the thickness of the paint coating must be governed by not more than the value cited by the American code (300 μm). Moreover, painting the mild, low tensile steel at full bond strength is prohibited due to friction; when it is painted with coating, all the bond strength will be lost, so it is important to avoid coating the smooth bars. Care must also be taken not to increase the thickness of the paint coating to more than 300 μm. Some researchers have stated that when a paint thickness of 350 μm was used for the main steel reinforcement in the concrete slab, testing found too many cracks that led to separation between steel bars and concrete.

Corrosion rates between steel bars without coating and other bars coated by epoxy were compared; both were exposed to tap water and then some samples were placed in water containing sodium chloride and sodium sulfate. Table 7.2 shows the rate of corrosion for the coated and uncoated steel bars. From this table, one can find that the corrosion rate is slower for coated than for uncoated steel bars. This method is cheap and is widely used by workers and contractors in North America and the Middle East.

Note that the coating of reinforcement steel by epoxy is not exempt from the use of concrete of high quality while maintaining a reasonably concrete cover. Some steel manufacturers can provide steel bars with the required coating. This is a good alternative to performing coating on-site because the thickness of the coating cannot be controlled as it needs some special tools to measure this thickness.

TABLE 7.2
Corrosion Rate for Coated and Uncoated Steel Bars

	Corrosion Rate (mm/Year)	
Case	**Tap Water**	**NaCl 1% + Na_2SO_4 0.5%**
Uncoated	0.0678	0.0980
Coated	0.0073	0.0130

7.4 GALVANIZED STEEL BARS

Some research in the United States has recommended using galvanized steel in reinforced concrete structures. Moreover, an FHWA report recommended that the age of galvanized steel reach up to 15 years when high-quality concrete is used and under the influence of chloride attacks (Andrade et al. 1994). Galvanized bars are used effectively in structures undergoing carbonation. Depletion of the galvanized bars accelerates if galvanized bars are mixed with nongalvanized bars.

The process of galvanization occurs through the use of a layer of zinc. To illustrate this briefly, galvanization immerses the steel rod in a zinc solution at a temperature of 450°C and then undergoes a process of cooling. The zinc cover is then formed on the steel bar. This cover consists of four layers: the outer layer is pure zinc and the other layers are a mix of zinc and steel. In zinc, corrosion will occur over time. The rate of corrosion under different weather conditions can be calculated by the corrosion on the zinc layer and the time required for it to corrode. Note that the relationship between the layer thickness and lifetime is represented by a linear relation, as illustrated in Figure 7.1. The galvanized coating must be tested after bending the steel bars. During welding in manufacturing, the maximum zinc cover formed is around 200 μm thick, following ASTM A767/A767A M-90.

The stability of zinc essentially depends on the stability of the pH value of concrete. A pH equal to 13.3 is the value at which the passive protection layer forms, but when the value of pH increases, the zinc will melt until it vanishes completely—the situation is illustrated in Figure 7.2. Therefore, the process of galvanization is totally dependent on the proportion of pH in the concrete pores and relies mainly on the alkalinity of portland cement. Hence, the shortage of cover when galvanization pH is equal to 12.6 will be 2 μm; in the case of pH equal to 13.2, shortage in the galvanized layer reaches 18 μm. This happens before the passive protection layer is present.

Laboratory tests have been conducted using different types of portland cement with different alkalinities. Assuming that corrosion is equal, when the thickness of the cover is equal to 60 μm, a lifetime of about 200 years can be expected when alkalinity is low and 11 years when it is high (Building Research Establishment 1969).

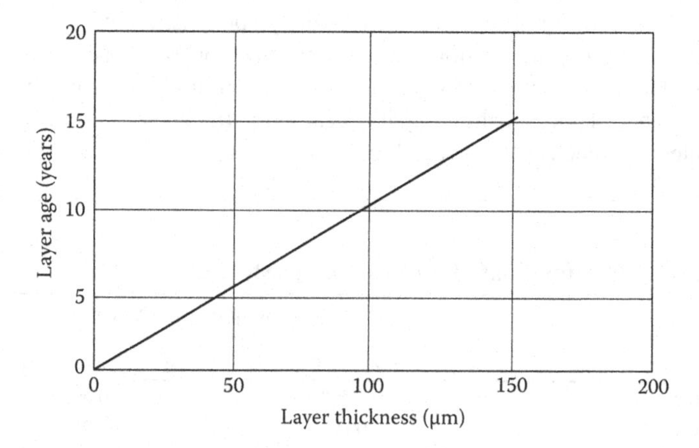

FIGURE 7.1 Relation between zinc layer thickness and expected lifetime.

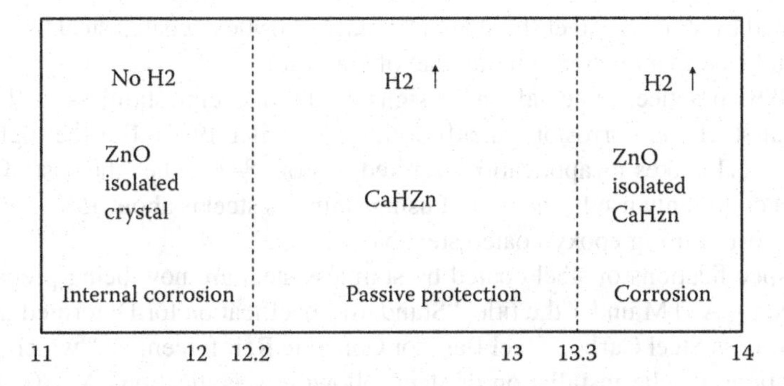

FIGURE 7.2 The galvanized steel behavior in solution with pH 11–14.

Therefore, the thickness of the cover must be more than 20 µm in the American specifications (ASTM 767/A767 M-90). This specification is especially for galvanized steel bars, for which two types of galvanization can be identified: first class I and type II class, which have cover thicknesses of more than 1070 and 610 g/m^2, respectively—equivalent to 85 and 150 µm, respectively.

The United Kingdom's Building Research Establishment (1969) recommended a maximum thickness of the zinc cover of about 200 µm; an increase in thickness of cover will reduce the bond between the steel and concrete. As stated before, when pH is between 8 and 12.6, zinc is more stable than for an increased pH value. The use of zinc provides high efficiency in the event that the structure undergoes significant carbonation, which reduces the pH, as we have discussed previously.

In the case of chloride attacks, galvanization will not prevent corrosion but will significantly reduce the rate of corrosion. Galvanization can increase the value of chlorides, which can increase corrosion to 150%–200%; this will increase the time to cause corrosion to four times longer (Yeomans 1994). The following are some of the important points that should be followed during galvanization, according to the specifications of the Institute of Equatorial Concrete in 1994:

- Galvanization increases the protection of steel corrosion but does not compensate for the use of concrete with an appropriate concrete cover.
- Galvanized steel should not be placed with nongalvanized steel as this will quickly cause corrosion of the galvanic layer.
- It is necessary to examine the galvanic layer after bending the steel reinforcement and fabrication and to increase the bending diameter.
- Some precautions should be taken when using welding with galvanized steel.

7.5 STAINLESS STEEL

Some very special applications use reinforcing steel bars made from stainless steel material in order to avoid corrosion. This process is more expensive, so normal steel bars coated by a layer of stainless steel 1–2 mm thick are used. Here, recall the same precautions that have been mentioned for galvanized steel: Stainless steel or steel bars coated by stainless steel should not be placed beside uncoated steel bars because this would lead to an increase in the rate of corrosion.

In a 1995 practice, some balcony designs used reinforcing stainless steel adjacent to normal steel and corrosion quickly followed (Miller 1994). But the high cost of stainless steel makes its applications limited—about 8–10 times the cost of the normal steel commonly used. The cost of using stainless steel is about 15%–50% higher than the cost of using epoxy-coated steel bars.

The specifications of steel coated by stainless steel are now being prepared and discussed by ASTM under the title, "Standard Specification for Deformed and Plain Clad Stainless Steel Carbon Steel Bars for Concrete Reinforcement," which provides specifications for the installation of steel following specifications A 480/480M for types 304, 316, or 316L. Three levels of the yield stress are defined: 300, 420, and 520MPa.

In the commercial market, reinforcing steel covered by stainless steel has a yield strength of about 500MPa, with a maximum tensile stress of about 700MPa; the available diameters are 16, 19, 22, 25, and 32mm. Some factories are now manufacturing 40mm diameters. It is not preferable to use stainless steel when welding is done, but, if necessary, welding can be done using tungsten energy; the welding wire will be of the same material as stainless steel.

It is found also that using stainless steel strands in prestressed concrete structure is excellent in extremely aggressive environment. As per Roddenberry and Al-Kaimakchi (2021), the benefits of using stainless steel strands include prolonged service life and fewer inspections of the structure. The flexural behavior of stainless steel prestressed concrete girders was experimentally studied. This study was performed on Seven full-scale 42ft (12.8m) long AASHTO Type II girders were designed, cast, and tested in flexure. Two of the seven girders had carbon steel strands and served as control girders. Experimental results showed that the overall flexural behavior of the girders prestressed with stainless steel strands is different from those prestressed with carbon steel strands. The capacity of all stainless steel girders increased up to failure, which reflects the stress-strain shape of the stainless steel strands. When the girders had the same initial prestressing force, the ultimate capacity of the stainless steel non-composite and composite girders was approximately 11.7% and 23.7% higher than that of the control girders, respectively. Experimental results revealed that regardless of failure mode, the girders prestressed with stainless steel strands can achieve ultimate capacity and deformability as high as those prestressed with carbon steel strands. Although the composite stainless steel girders failed due to rupturing of strands, they failed at a noticeable deflection with many flexural cracks in the midspan. Rupture of strands failure mode is particularly important because it demonstrates the importance of the ultimate strand strain in design. The guaranteed ultimate strain of stainless steel strands is 1.4%.

There was a recent study performed by Wang et al. (2022) by doing a comparison of the corrosion behavior of corrosion-resistant steel reinforcements, including epoxy-coated steel, high-chromium steel, and stainless steel reinforcement in normal-strength concrete and high-performance concrete (HPC) columns in an accelerated chloride attack environment for 2 years. The corrosion potential and corrosion rate of the reinforcements were monitored using electrochemical methods, and the degradation of the axial compressive capacity of 40 corroded columns. Findings indicated that corrosion-resistant reinforcements showed significantly better corrosion performance: no corrosion was observed for intact epoxy-coated and stainless steel reinforcements, and less corrosion (54%) was found on high-chromium steel than conventional mild steel in normal concrete, while similar corrosion rates were found for mild steel and high-chromium steel reinforcements in HPC. Results also indicated that HPC provided reliable protection to the embedded reinforcements, showing smaller corrosion rates than those in normal concrete. The measured average corrosion rate of mild steel and high-chromium steel reinforcements in HPC was 17%–37% of that in normal concrete.

7.6 FIBER REINFORCEMENT BARS

In the last decade, different theoretical and practical research studies have been conducted on the replacement of steel reinforcement bars by using filler-reinforced polymer (FRP) because this material is not affected by corrosion and thus will be very economical during the course of a structure's lifetime. Figure 7.3 shows that the same shape of rolled steel section made from GFRP (glass-fiber-reinforced polymer) is more expensive than the traditional steel sections. On the other hand, there will be no cost of painting and maintenance work as they are not affected by corrosion under any environmental conditions surrounding the structure. Moreover, it is lightweight as the density of the fiber is 2.5 g/cm^3; the density of steel is about 7.8 g/cm^3, so its weight is three times less than that of steel.

Note that its weight is very low and its resistance is more than the resistance of steel, so it will be worth using for the addition of more floors to an existing building. Currently, these sections are used to replace normal steel grating, handrails, and ladders in offshore structure platforms in the oil industry. These bars have been developed and manufactured from these materials as alternatives to steel bars. Figure 7.4

FIGURE 7.3 GFRP structural sections.

FIGURE 7.4 Shape of the reinforcement bar from GFRP.

shows that manufacture takes place in such a way that protrusions occur on the surface, which gives the ability to form bonds between them and the concrete. Currently, the manufacture and use are on a small scale as they are under experimentation in the market. Much research is being conducted to study the performance of these bars with time. As a result of limited production, the cost is very high; however, the maintenance is very low or can be ignored. Moreover, the value of the dead load on the structure will be reduced, which will reduce the total cost of the structure.

Some structures in Canada, such as marine structures in ports, or wharfs, have been established using bars made of GFRP. These structures are composed of thick walls of precast concrete designed to have a maximum strength of 450 kg/cm^2. They are exposed to temperatures ranging from 35°C down to –35°C. Bridges in Quebec, Antherino, and Vancouver Island, Canada, for which salt has been used to dissolve ice, have been built using the same kind of reinforcement. Tests have been conducted on all these facilities through sampling; the reinforced concrete has been studied using an X-ray machine. These structures are aged between 5 and 8 years. The research found that various factors, including wet and dry cycles, had no impact on the GFRP.

Microscopic examination found strong bonds between the bars manufactured from GFRP and concrete. The maximum strength was about 5975 kg/cm^2 and the maximum bond strength was 118 kg/cm^2; the modulus of elasticity of GFRP was about five times less than the steel modulus of elasticity. To overcome such problems, which might arise from the creep, the American specification ACI 440 has suggested that tensile stress must be not less than 20% of the maximum tensile strength.

7.7 PROTECTING THE CONCRETE SURFACE

A concrete surface needs to be protected from permeability of water and propagation of chloride—for example, offshore structures exposed to sea waves, or bridge decks that use salt to melt ice; in addition, permeability needs to be prevented in order to protect concrete surfaces from carbonation and inhibit propagation of carbon dioxide.

The other factors that increase the corrosion rate—humidity and oxygen—need to be prevented from spreading inside the concrete so that the corrosion rate will decrease. There are two popular methods of protection of concrete surfaces: spraying liquid materials or painting by brush and using sheets and membrane from rubber and plastic or textiles immersed in bitumen, which is usually used in small projects.

This isolation from water has the same problems for painting epoxy on reinforced steel as we discussed before, such as preventing any damage to this layer or erosion due to external mechanical factors such as tearing or existing heat sources and other factors. Thus, it is important to follow the required specifications during execution. Some specifications define the ways of execution in addition to the supplier's recommendations. There are different types of surface protection using liquid materials, but the function is the same: To prevent the propagation of water inside concrete through concrete voids, as shown in Figure 7.5.

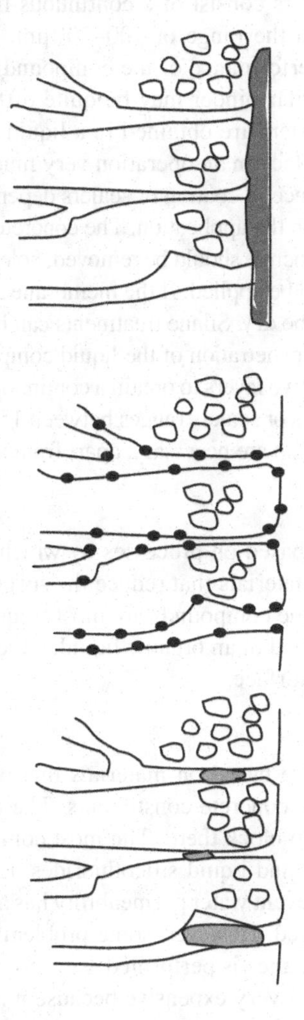

FIGURE 7.5 Sketch presenting different coatings.

7.7.1 Sealers and Membranes

Sealers and membranes have been used traditionally for providing protection to concrete structures exposed to severe chemical attack. However, with growth in the frequency of durability problems generated by corrosion of steel in regular reinforced concrete structures, they are increasingly used as a means to mitigate this durability problem. Membranes and sealers can provide protection by (1) eliminating or slowing down the penetration of chlorides and carbonation, to keep the steel passivity; and (2) reducing moisture movement into the concrete to keep it dry and slow the spread of corrosion reactions. The sealers and membranes can be classified into several types, and each type represents a family of materials with different chemical compositions.

7.7.1.1 Coating and Sealing

Coating materials and sealers consist of a continuous film applied on the concrete surface with a thickness in the range of 100–300 μm. The film is composed of a binder and fillers for the performance of the compounding of the fillers. Therefore, compositions having a similar binder may be quite different in their performance. The coating material or sealers are obtained in a liquid form, which are brushed or sprayed on the concrete surface in an operation very much like painting.

The successful performance of coating or sealers depends not only on the quality of the materials used but also on the application. The concrete surface should be clean and sound. Weak and cracked concrete should be removed, holes should be filled, and, if necessary, a leveling coat should be applied. If the membrane applied has a polymeric composition, the surface should be dry. Silane treatments can be done in wet concrete. This is needed to facilitate better penetration of the liquid compound into the pores. Coating should usually be applied in two layers to obtain a continuous film without pinholes. The service life of the membranes or sealers ranges between 15 and 20 years. Thus, continuous maintenance and follow-up are necessary, apart from additional treatments.

7.7.1.2 Pore Lining

Pore lining treatments are based on processes by which the surface of the pores in the concrete is lined with materials that reduce the surface energy to make the concrete water-repellent. Silicone compounds are most frequently used for this purpose. Silicone resin can be dissolved in an organic liquid, which, after evaporation, deposits a resin film on the pore surface.

7.7.1.3 Pore Blocking

Pore blocking treatments are based on materials that penetrate into the pores and then react with some of the concrete constituents. The resulting insoluble products are deposited in the pores to block them. The most common materials used for this purpose are liquid silicates and liquid silicofluorides. Experience has shown that a membrane that is used to prevent water permeability has a lifetime of around 15 years; after that, it must be replaced. However, some problems occur (e.g., in bathrooms) when this proactive maintenance is performed.

The proactive approach is very expensive because it needs to remove all the tiles and plumbing accessories and start from scratch; therefore, in residential buildings, the maintenance will be corrective. In any mature reinforced concrete structure, most of the problems due to isolation will be found because the main steel of the slab is lower and the signs of corrosion will be parallel cracks to the steel bars in the main steel bars. Such signs will be seen on the lower floor, so, for example, a user of a bathroom cannot observe this. Therefore, the repair will be done until the concrete cover falls and serious corrosion problems occur in the steel bars. When the membrane is used, one must take care in execution as any tear in the membrane or bad execution that does not follow the specifications and supplier recommendations will lead to loss of money. Any defect will cause permeability of water and start corrosion.

There are two well-known examples of isolation for foundations: swimming pools and tanks. Here, methods isolate the surface that is exposed to water or isolate the surface that is not exposed directly to water. This type of isolation must be executed by a professional contractor experienced in this type of work because using the chemicals and dealing with the membrane need competent workers available only from a trustworthy, specialized contractor. Recently, research has found that problems occur in isolation using a membrane in cases of cathodic protection because some gases accumulate in the anode and need to escape; however, the membrane will prevent this and thus prevent completing the electrical circuit.

7.7.2 Cathodic Protection by Surface Painting

The corrosion of steel in concrete occurs as a result of chloride attacks or carbonation of a concrete surface and incursion into the concrete until it reaches steel, which reduces concrete alkalinity. The presence of moisture and oxygen brings about practical corrosion and will continue until a complete deterioration of concrete takes place, as has been clarified in Chapter 2. Some materials are used to paint the surface of concrete to a saturation point. Then this material will spread through the concrete to reach steel at speeds of 2.5–20 mm per day up to the surface of steel by capillary rise, such as water movement; by penetrating concrete with water in cases of chloride attack; or through propagation by gas such as carbon dioxide when exposed to surface carbonation.

Therefore, when these materials reach the surface of steel, an isolation layer around the steel bar's surface will reduce the oxygen in the cathode area to the surface on the cathode and reduce the melting of steel in the water in the anode area, thereby delaying the process of corrosion and reducing its rate. Figure 7.6 presents the influence of this cathodic protective coating on the surface in protecting the steel bars. These new advanced materials are used for new construction or for existing structures in which corrosion has started in steel bars. They are also used when corrosion is clearly present and complete repair to the damaged concrete surface is required followed by painting of the surface.

7.8 CATHODIC PROTECTION SYSTEM

This method is the most expensive method of protection; it is usually used in protecting pipelines in the petroleum industry. It is intended for use in reinforced concrete structures and also for special structures due to its higher cost and need for special

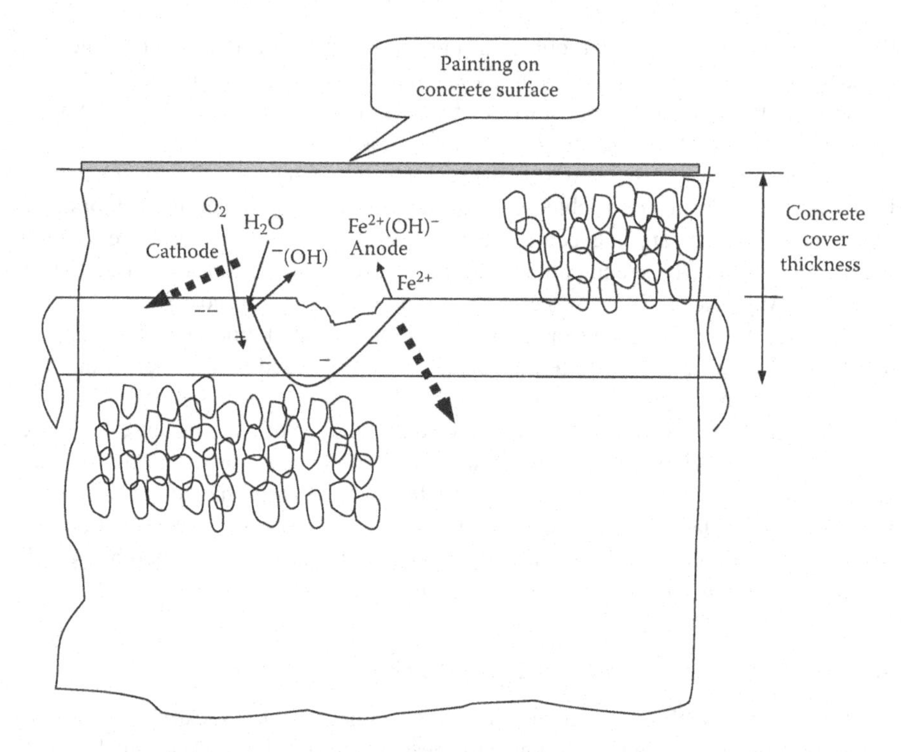

FIGURE 7.6 Painting a concrete surface to provide cathodic protection.

studies, design, execution, and monitoring. If applied properly, cathodic protection can prevent corrosion of steel in concrete and stop corrosion that is already in progress. This is accomplished by making the steel bar a cathode by using an external anode. Electrons are supplied to the reinforcing bar from the anode through the ionically conductive concrete. The current supplied should be sufficiently high so that all the local cells are inhibited and the entire steel surface becomes anodic. The external current can be supplied by connecting the steel to a metal that is higher in the electrochemical series (e.g., zinc). This serves as the anode relative to the cathodic steel. In this method, the anode gradually dissolves as it oxidizes and supplies electrons to the cathodic steel. This type of cathodic protection is called a "sacrificial" anode protection.

An alternative method for cathodic protection is based on supplying electrons to the reinforcing steel from an external electrical power source. The electrical power is fed into an inert material, which serves as the anode and is placed on the concrete surface. This method is referred to as "impressed current anodic protection." The anode is frequently called a "fixed" anode.

7.8.1 Cathodic Protection

The principle of the use of a sacrificial anode was put forth in 1824 by Sir Humphrey Davy. Afterward, the discoveries were used for boats completely submerged under water to protect their metal parts from corrosion. At the beginning of the twentieth century, the technology was used for underground pipelines, but when it was discovered that the soil's resistance to electricity was very high, cathodic protection with the current and constant direct current was used. Cathodic protection is used in the construction of special modern structures having more importance. The first practical application in the area of reinforced concrete was a bridge on a highway in the mountainous area north of Italy.

The use of cathode protection became more widespread with the development of research and technology. As Broomfield (1995) stated in his survey of consultants working in the area of cathode protection of concrete structures in England, protection of cathodic disabilities has been established for about 64,000 m^2 in about 24 structures in the United Kingdom and the Middle East. The largest manufacturer of anodes has supplied anodes that cover about 400,000 m^2 around the world and include a garage for 60 cars and 400 different forms of bridges and tunnels.

Cathodic protection is used in a certain type of construction as a result of the high cost as well as the specific nature of the construction and the need for a system of monitoring. It is used most often in structures exposed to chlorides that exist inside the concrete mix, or penetration of chlorides within the concrete resulting from surrounding environmental conditions. Chloride influence in the presence of concrete has a special nature in that it starts when corrosion results from chlorides and the part of the concrete contaminated by chlorides must be completely removed.

Sometimes, the work of repair is very difficult. The case of chlorides in the concrete mix is considered impossible. Experience shows that the limited availability of electrical protection is more effective in stopping the process of corrosion, as mentioned in some previous studies, than the traditional way in the case of concrete pollution by chloride. The FHWA in the United States has said that the method of repair proven to be the only way to stop corrosion in concrete bridges with salts is the use of electrical current, irrespective of the chloride content in concrete.

The use of cathode protection specifically prevents the corrosion of steel reinforcement in concrete or stops the corrosion process when it has already begun; this method relies on making steel with a continuous cathode through the use of an external anode (Figure 7.7). As seen from this figure, when cathodic protection is not used, a cathode with a negative electron will form on the surface of a steel part and the other part will work as an anode; from this, the corrosion process forms, as explained in Chapter 4. As shown in Figure 7.7, the electron will be generated on the steel reinforcement surface when an anode is placed on the concrete surface. In this case, there exists electrical conductivity between the concrete and the steel reinforcement, so the cathodic protection will be formed on the steel reinforcement. As shown in the figure, the positive electrons will move to the anode. This method of cathodic protection is called sacrificial protection and the anode in this case is called a sacrificial anode.

There are two main methods of cathodic protection. The first depends on using a sacrificial anode, and the method is called sacrificial protection. In this case, the anode will be made of zinc metal that will get corroded instead of the steel reinforcement. The oxidation process will cause the zinc to move to the steel reinforcement and the negative electron will be formed on it; this is the required cathodic protection.

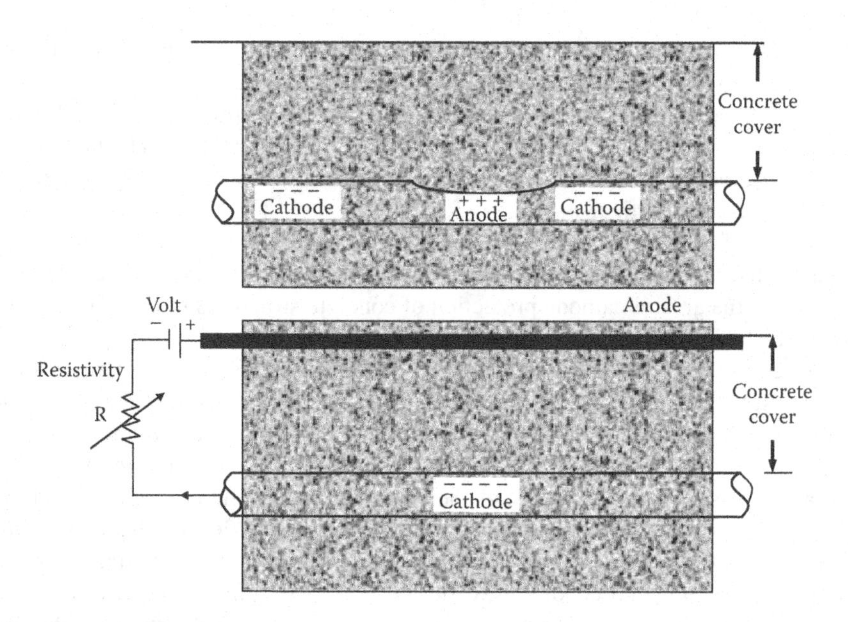

FIGURE 7.7 Principle of CP method.

The second method of cathodic protection will generate electrons on the steel reinforcement in concrete by outsourced electricity. This source will be the anode and will be placed inside the concrete and called a fixed anode. Sacrificial protection is used in submerged structures; the concrete is immersed in water so that there will be little electron movement and the potential voltage between the two materials will be small. This will provide cathodic protection for a long time.

Figure 7.8 illustrates the second method of protection—putting an anode in concrete and using an external source of electricity. An example of a fixed anode is a wire mesh, placed in concrete at the concrete surface; it works as an anode; the conductivity among anodes, steel reinforcement, and batteries by using cables is shown in Figure 7.8. The source of energy is normally used batteries. A RILEM report put some consideration into more valuable cathodic protection:

- The electrical conductivity to the steel reinforcement must be continuous.
- The concrete between steel reinforcement and anode must be able to conduct electricity.
- Alkaline aggregates must be avoided.

7.8.2 Cathodic Protection Components and Design Consideration

The cathodic protection system consists of impressed current, anode, and the electric conductive element, which is the concrete in this case. It depends on the relative humidity of concrete as it has a great effect on the electrical cables and the negative pole system on the steel reinforcement. The last component of this system is the wiring that will conduct between the anode and the source of direct current. But we must not forget the last important element, which is the control and measurement

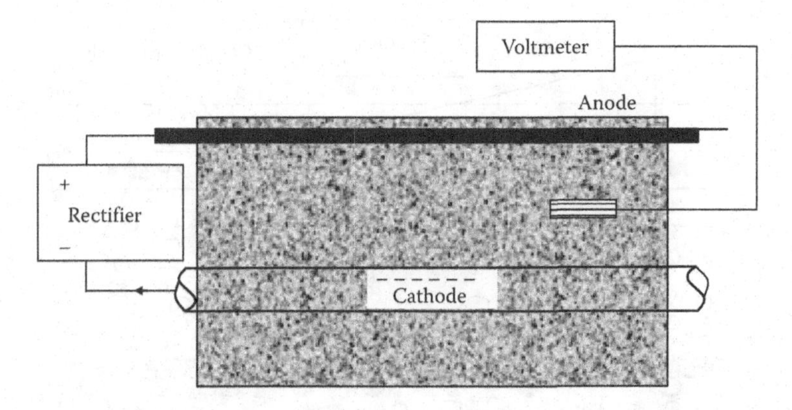

FIGURE 7.8 Cathodic protection method.

system. The most important and expensive element in the cathodic protection system is the anode, as it should be able to resist the chemical, mechanical, and different environmental conditions that affect it over the structure's lifetime. In general, its long lifetime is preferable to that of its coating layer life. The anode surface must be large and have a little density to the electric current so as to prevent any deterioration of the anode.

From a previous discussion, it is clear that cathodic protection of a concrete structure is different from any other applications because, in some parts, concrete has pores that avoid containing water and other parts may be dry or have water on the surface only; thus, marine structures and structures buried in soil are different. The main fact is that the concrete structure cannot contain 100% water inside its pores. This is clear if the concrete around the anode is completely dry; the potential volts need to be increased by using 10–15 V instead of 1–5 V. On the other hand, the dry condition around the steel bars prevents causes of corrosion. Practically speaking, this cannot happen and only can be seen in summer without high relative humidity.

7.8.2.1 Source of Impressed Current

Most design methods take a current of about 10–20 mm ampere on the surface meter for steel reinforcement and take into consideration the lower layer of steel reinforcement. Potential voltage of around 12–24 V is always used to ensure that humans or animals experience little electric shock. The source of electric current is chosen in the design stage; the choice of electric source will be based on the fact that it must be enough to stop the corrosion process. A higher estimate for electric current is often assumed and 50% is added as a safety factor. This increase in current will produce little heat that will lose some part of the electric current.

7.8.2.2 Anode System

The most important element in the cathodic protection system is the anode. As the anode is the most expensive element of the system, experience in choosing and installing it is also needed. The choice of the anode system depends on the type of the structure, its shapes, and other requirements. There are two main types of anodes:

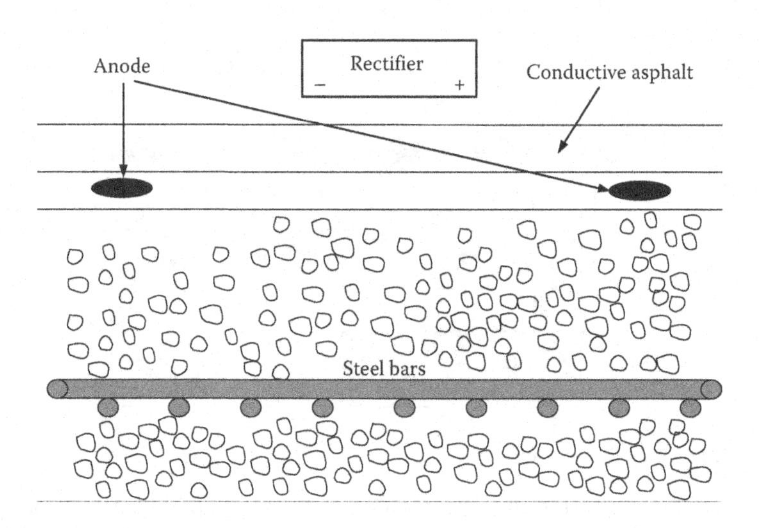

FIGURE 7.9 Sketch for using conductive asphalt.

The first type is used on bridge decks and fixes the deck from the top surface; it needs special properties to accommodate moving vehicles and will be covered by a layer of asphalt. The second type is used in vertical buildings; these anodes do not need high resistance to abrasion, as the first type does.

Anodes for bridge decks. From an execution point of view, there are two types: The first buries the anode in the concrete top layer, as shown in Figure 7.9. The main problem with this type is that it increases the covering layer that will coat the anode, which consequently will increase the dead load on the structure. Therefore, this is not an economic solution. The second executable type is by making holes on the concrete surface and installing the anode inside. This method can overcome the increase of dead load on the structure more easily than the first type. On the other hand, it will be difficult to cut parts of the concrete as the anodes should be adjacent to each other by about 300 mm to maintain the suitable distribution of electricity. The first type can use carbon anodes or silicon iron with 300 mm diameter, 10 mm thick discs buried inside asphalt coke; the upper layer is an asphalt conductive layer as a secondary anode. The second type, as shown in Figure 7.10, makes 300 mm diameter holes on the upper concrete surface; the main target is to reduce the dead load on the bridge, consequently reducing the overall construction cost and reducing the increase of the bridge surface leveling.

Vertical surface anodes. In this type of concrete structure member, titanium mesh is always used and often concrete is applied by using shotcrete on the surface of the mesh to cover the anode. Strong precautions are needed in using shotcrete as it needs more competent workers, supervision, and contractors. It also needs special curing to guarantee good-quality concrete and good adhesion between it and the old concrete surface. Some precautions must be considered during construction to ensure that the layer saturated 100% will bond with old concrete and the electric conductivity will be through pieces of titanium sheet that will weld on the mesh, as shown in Figure 7.11.

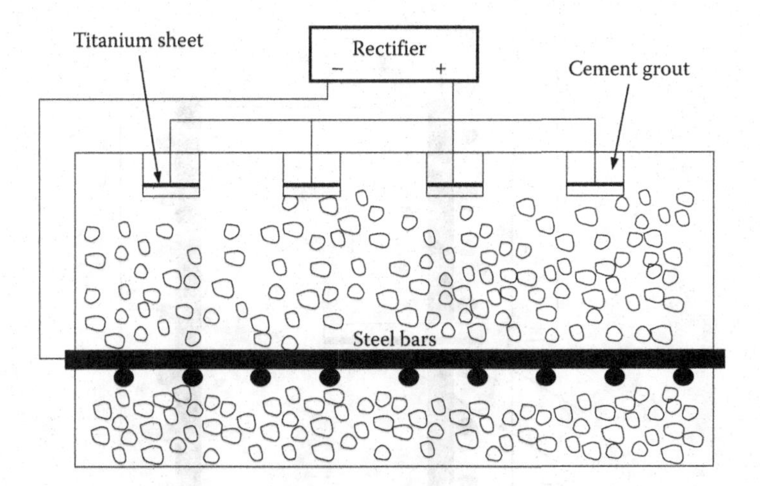

FIGURE 7.10 Sketch presents anode holes at the surface.

FIGURE 7.11 Fixing titanium on a concrete column.

We may use the rod of titanium platinum placed inside the concrete member of a subject inside the coke so as to reduce the impact of acid generated on the anode. One of the most important precautions to be taken during implementation is to make sure that there is no possibility of causing a short circuit between anode and steel reinforcement. To prevent that, making the use of an appropriate cover, after making a hole in the concrete, will give a warning when it comes into contact with steel bars, as in Figure 7.12. The anodes should be distributed by design system in order to ensure the protection of steel reinforcement in the structure. This method is also used in horizontal surfaces.

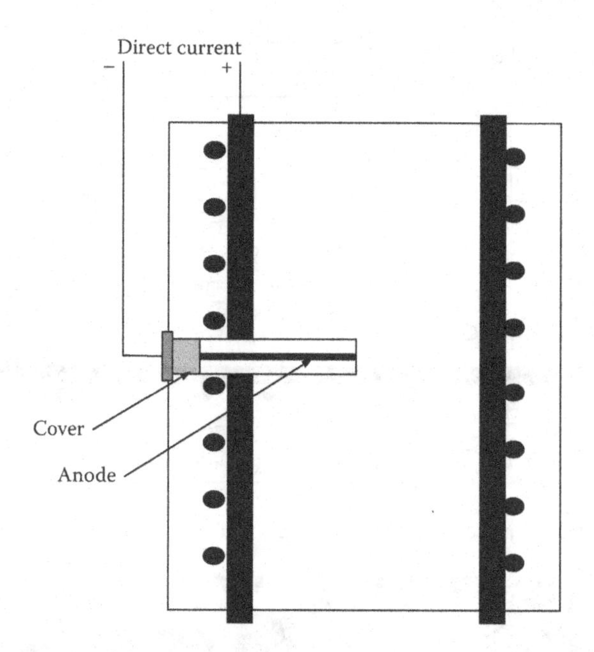

FIGURE 7.12 Sketch represents a buried anode.

7.8.2.3 Conductive Layer

The anode is a main pole of inert metal; the secondary poles consist of layers from mortar, asphalt, and coating. This layer conducts the electric current by the carbon particles. Figure 7.13 shows the conductive layers in the case of vertical reinforced concrete structures like columns and walls. These layers do not have the capability to be durable with time, as the titanium mesh does, but are considered the most inexpensive and easiest way from a construction and maintenance point of view. They can be painted by a reasonable painting match with the architectural and decoration designs for the building, ceiling, and vertical and horizontal surfaces. The main disadvantage is that they cannot be used in a surface exposed to abrasion as capacity is low and their lifetime is about 5–10 years.

7.8.2.4 Precautions in Designing the Anode

The most important and most expensive part of a cathodic protection system is the anode. Therefore, some precautions and specifications are required, including selection of a type appropriate for the structure; every type has advantages and disadvantages, so it needs to be studied based on the application. As a result of the chemical reactions, the alkalinity of the concrete on the surface decreases, increasing the acidity. The material using the anode may have a high resistance to acid, but the surrounding concrete may not have this capability, so it must be considered as the amount of accumulated acid proportional to the density of the electric current. As a result of the chemical reactions, the alkalinity of the concrete on the surface decreases, consequently increasing the acidity in the concrete. In the case of using shotcrete with cement to aggregate, the ratio is 1:5; in the case of increasing the current density, the deterioration rate will increase in addition to that for existing chlorides, and

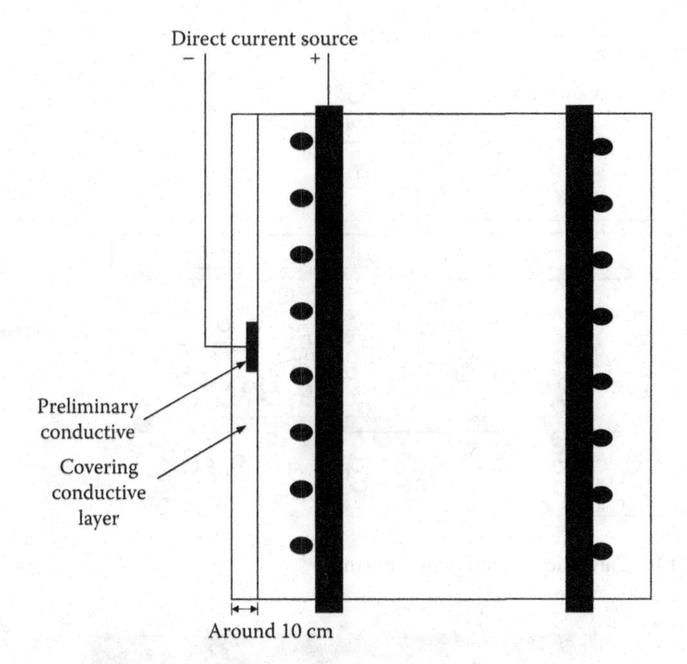

FIGURE 7.13 Present anode conductive layer.

increasing it has an effect on the increasing acid rate. Therefore, increasing the area of anode to concrete surface area will be better from anode consumption and the acid formation.

7.8.2.5 Follow-Up Precaution

Note that the cathodic protection needs to be checked periodically to ensure that the cathodic protection system works well without any obstacles, as in Figure 7.14. The cell pole near the steel reinforcement (see Figure) can be connected easily by a microprocessor via a cable modem, which can measure and transfer data to the computer directly. Consequently, this system can work without having to make periodic visits to the site. The computer, which has special software and hardware, can monitor more than one point in the structure. This computer can monitor the efficiency of the cathodic protection. Advanced systems can be used to control the source of the electric current by changing the current density and potential from a long distance by using the microprocessor.

7.8.3 Comparison between Cathodic Protection and Other Types of Protection

Cathodic protection needs to be measured with each period of time or by measuring away from the structure, as we have said before, using the computer to ensure the safety of the design and the proper functioning of the cathodic protection system. Therefore, cathodic protection is a costlier technique compared to other types of steel reinforcement protection and is often used in important structures with a special

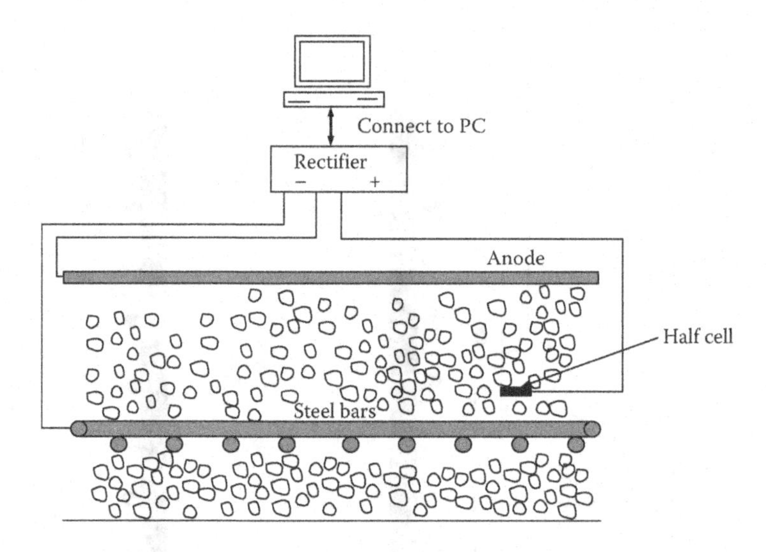

FIGURE 7.14 Cathodic protection monitoring.

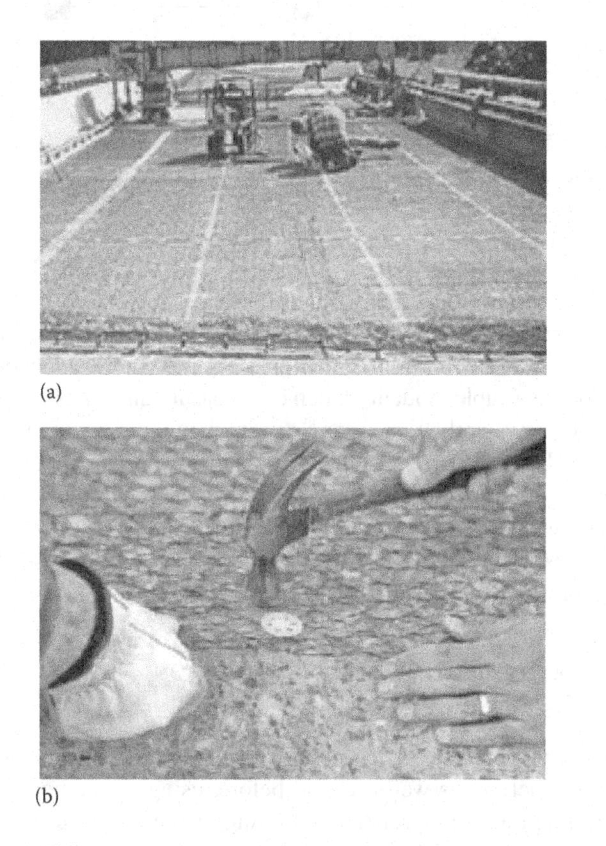

FIGURE 7.15 (a) Fixing an anode on a bridge deck and (b) fixing titanium mesh using a plastic binder.

FIGURE 7.16 The installation of titanium bars with steel bars.

nature. Hence, it has been used in bridges and tunnels since the late 1980s and early 1990s, including an underground garage at the World Trade Center in New York City, as well as structures in the United Arab Emirates. Kendell and Daily (1999) reviewed the implementation method of cathodic protection during the implementation of the civil works, noting that it does not need excess labor for implementation as anodes are placed with reinforced steel (Figure 7.15a). Figure 7.15b shows the development of a titanium mesh with reinforcing steel and how to stabilize the titanium mesh through a link from the plastic. Figure 7.16 shows titanium bars; this method does not need extra workers as the workers that place the steel reinforcement will also place the titanium.

7.8.4 Cathodic Protection for Prestressed Concrete

Most specifications state that cathodic protection must not be used when prestressed concrete is used. As was explained before, the movement of electrons forms hydrogen ions on the surface of a steel reinforcement. The presence of hydrogen affects composition of iron atoms for the high-strength steel that is usually used in the case of pretension of the steel in prestressed concrete. Cathodic protection can be used, taking into account that hydrogen will not affect the nature of iron and the quantity of hydrogen resulting from the cathodic protection technique does not pose a threat to the prestressed steel stress. Some prestressed structures do not use cathodic protection of underground structures. In some cases, in Florida, sacrificial anodes have been used in bridge pilings by using depleted anodes (Kessler, Powers, and Lasa 1995).

7.8.5 Bond Strength in Cathodic Protection

The impact of electric current on the bond strength between steel reinforcement and concrete has been considered to be one of the most important subjects studied by researchers for several years. From a practical point of view, there are no notes for any impact of cathodic protection on bond strength. Some structures that have been studied are about 20 years old and there has been little impact on the bond strength, as most of it is carried out by the ribs that exist in the deformed steel reinforcement.

Some practical experience has proved that bond strength increases when rust is present on the steel surface. When cathodic protection is used, there is no need to use another protection method, such as membrane, which prevents water permeability, or painting the steel bars, and so on. This reduces the total cost of the project and, for that, in deteriorating concrete structure or some of its members, it is used to stop more deterioration of the structures but it will be a high cost so it can be used if it is impossible to perform a normal repair. Therefore, the cost of repairing the concrete member may be higher than the cost of constructing it.

In some structures, due to the special nature of construction, it is difficult to work the necessary repairs, so the cathodic protection is the best and only solution. Table 7.3 contains a summary of the methods of protection prepared by Kendell and Daily (1999); it gives the advantages and disadvantages of each method and the appropriateness of use of cathodic protection at the same time.

7.9 CONCRETE WITH SILICA FUME

Concrete is generally classified as normal-strength concrete (NSC) and high-strength concrete (HSC), but there is no clear-cut boundary between the two classifications. International codes generally define HSC as concrete with strength equal to or greater than 50 N/mm^2. Recently, a new term, "high-performance concrete" (HPC), has been used for concrete mixtures that provide concrete with high workability, high strength, high modulus of elasticity, high density, low permeability, and resistance to chemical attack.

There is a slight controversy between the terms "high-strength" and "high-performance." HPC is also HSC, but it has a few more attributes specifically designed, as mentioned earlier. But, in terms of concrete durability, HPC is very important as its permeability is very low. Therefore, the spread of carbon dioxide, oxygen, or chloride will be very small, which will enhance the durability of the concrete from corrosion.

The main process to obtain HPC is to reduce w/c ratio to less than 0.3, which will greatly improve quality expected from the HPC. The main approach for obtaining high workability and high strength is the use of silica fumes, which become a

TABLE 7.3
Advantages and Disadvantages of Different Types of Protection

Method	Advantage	Disadvantage	Compatible with Cathodic Protection?
Increase concrete cover	No cost increase Obtain from good design	Has limits Increasing cover will increase the shrinkage crack	Yes
Not permeable concrete	Depend on design	Need additives with very good curing	Yes, but silica fume must be <10% cement weight
Penetrating sealer	Relatively no cost	Complicated practically	Yes
Membrane prevents water	Very-well-known technology	Can be damaged Difficult to fix without fault	Yes, but cathodic protection under the membrane
Epoxy-coated steel bars	Well-known technology Little maintenance	Problem in quality control under certain weather conditions	Yes, with continuous electric current
Galvanized steel	Easy use on-site	Quickly damaged by contact with nongalvanized steel	Yes; need higher level of protection
Stainless steel	Excellent for corrosion prevention	High cost	Yes
Corrosion inhibitor	Calcium nitrate effect for little chloride concentration	Must know the chloride amount Effect over a long time not known	Yes
Cathodic protection	Well-known technology Long life by titanium anode	Regular inspection of rectifier	

necessary ingredient for strength above 80 N/mm^2. The ACI defines silica fume as a "very fine noncrystalline silica produced in an electric furnace." Also referred to as microsilica or condensed silica fume, this product results from reduction of high-purity quartz with coal in an electric arc furnace during the manufacture of silicon or ferrosilicon alloy. Silica fumes rise as an oxide vapor. Condensed silica fume is essentially silicon dioxide (more than 90%) in the noncrystalline form. They are considered to be an airborne material with a spherical shape. It is extremely fine, with particle sizes of less than 1 μm and an average diameter of about 0.1 μm, so it is about 100 times smaller than average cement particles. Silica fume has a specific surface area of about 20,000 m^2/kg; cement has a surface area from 230 to 300 m^2/kg. Due to its very small size, silica fume can close any pore void in concrete, so it enhances the properties of concrete by high strength and low permeability—the two main keys to having durable materials. Note that silica fume is more expensive than normal concrete strength due to the cost of silica fume and superplasticizer.

The corrosion behavior in ultra-high performance concrete in harsh marine environment was studied by Moffat et al. (2020). Their study revealed that the durability performance of ultra-high-performance concrete (UHPC) exposed to a marine environment for up to 21 years. Concrete specimens (152 x 152 x 533 mm [6 x 6 x 21 in.]) were cast using a water-cement ratio (w/cm) in the range of 0.09–0.19, various types and lengths of steel fibers, and the presence of conventional steel reinforcement bars in select mixtures. Laboratory testing included taking cores from each block and determining the existing chloride profile, compressive strength, electrochemical corrosion monitoring, and microstructural evaluation. Regardless of curing treatment and w/cm, the results revealed that UHPC exhibits significantly enhanced durability performance compared with typical HPC and normal concretes. UHPC prisms exhibited minimal surface damage after being exposed to a harsh marine environment for up to 21 years. Chloride profiles revealed penetration to a depth of approximately 10 mm (0.39 in.) regardless of exposure duration. Electrochemical corrosion monitoring also showed passivity for reinforcement at a cover depth of 25 mm (1 in.) following 20 years.

REFERENCES

ACI Committee 318. 1988. Revisions to building code requirements for reinforced concrete. *American Concrete Institute Structures Journal 85*(*6*):645–674.

Andrade, C., J. D. Holst, U. Nurnberger, J. J. Whiteley, and N. Woodman. 1994. Protection system for reinforcement. Task Group 7/8 of Permanent Commisson VII CEB.

Bentur, A., S. Diamond, and N. S. Berke. 1998. *Steel Corrosion in Concrete: Fundamentals in Civil Engineering Practice*. London, UK: E & FN Spon.

Broomfield, J. P. 1995. *Corrosion of Steel in Concrete: Understanding, Investigation and Repair*. London, UK: E & FN Spon.

Building Research Establishment. 1969. Zinc-coated reinforcement for concrete. *BRE Digest 109*.

Cairns, J. 1992. Design of concrete structures using fusion-bonded epoxy-coated reinforcement. *Proceedings of the Institute of Civil Engineering Structures and Buildings 4*(*2*):93–102.

El-Reedy, M. A., M. A. Sirag, and F. El-Hakim. 1995. Predicting bond strength of coated and uncoated steel bars using analytical model. MSc thesis. Giza, Egypt: Cairo University.

Gustafson, D. P. and T. L. Neff. 1994. Epoxy-coated rebar, handled with care. *Concrete Construction 39*(*4*):356–369.

Kendell, K. and F. S. Daily. 1999. Cathodic protection for new concrete. *Concrete International Magazine 21*(*6*):32–36.

Kessler, R. J., R. G. Powers, and I. R. Lasa. 1995. *Update on Sacrificial Anode Cathodic Protection of Steel Reinforced Concrete Structures in Seawater. Corrosion 95, Paper 516*. Houston, TX: NACE International.

Miller, J. B. 1994. Structural aspects of high-powered electrochemical treatment of reinforced concrete. In *Corrosion Protection of Steel in Concrete*, pp. 1400–1514, ed. R. N. Swamy. Sheffield, England: Sheffield Academic Press.

Moffat, E. G., Thomas, M.D., Fahim, A., and Moser, D.R., 2020. Performance of ultra-high-performance concrete in harsh marine environment for 21 years' *ACI Materials Journal 117(5)*:105–112.

Newston, C. and I. Robertson. (2019). Improving concrete durability through use of corrosion inhibitors. *ACI Materials Journal 116(5)*:149–160.

Roddenberry, R., and A. Al-Kaimakchi. 2021. Flexural behaviour of concrete bridge girders presresssed with stainless steel strands. *ACI Materials Journal 118*:137–152.

Pike, R. G. et al. 1972. Nonmetallic coatings for concrete reinforcing bars. *Public Roads 37(5)*:185–197.

Satake, J., M. Kamakura, K. Shirakawa, N. Mikami, and R. N. Swamy. 1983. Long-term resistance of epoxy-coated reinforcing bars. In *Corrosion of Reinforcement in Concrete Construction*, pp. 357–377, ed. A. P. Crane. London, UK: The Society of Chemical Industry/Ellis Horwood Ltd.

Wang., B., Belasrbi, A., Dawood, M., and Kahraman. R., 2022, Corrosion behaviour of corrosion resistant steel reinforcements in normal strength and high performance concrete: Large sacle column tests and analysis. *ACI Materials Journal 119*:89–102.

Yeomans, S. R. 1994. Performance of black, galvanized and epoxy-coated reinforcing steels in chloride-contaminated concrete. *Corrosion 50(1)*:72–81.

8 Repair of Reinforced Concrete Structures

8.1 INTRODUCTION

In most cases, repairing and retrofitting a concrete structure are much more complicated than building a new structure because new constructions do not need competent engineers with a lot of experience, but rather can be handled by junior engineers with a reasonable amount of experience (2 or 3 years). On the other hand, repair of concrete structures is more challenging and needs competent engineers and consultants as well. The building already exists and it is necessary to define new solutions to the problem that must be matched with the nature of the building and owner requirements and satisfy safety and economic requirements at the same time. The repair process for reinforced concrete structures is dangerous and hence very important; extensive care must be taken in choosing the suitable repair methods and tools.

The reasons for structure deterioration can be summarized as follows:

- Code and specification precautions were not considered in the design phase.
- Code and specification precautions were not considered in the execution phase.
- A suitable protection method was not selected in the design phase.
- A suitable protection method was selected, but the execution was bad.
- No maintenance monitoring was performed.

In all these cases, the result will be the same: structure deterioration, which will be obvious from the cracks in and concrete cover falling from the different reinforced concrete elements in the structure. This situation is bad. Shortages occur in a cross-sectional area of the concrete member, which requires speeding up the repair work to avoid worsening the condition of the structure, most likely leading to collapse with time. The first and most important step for any repair is to determine accurately what needs to be repaired; this depends entirely on the assessment of the structure and answers to the following questions:

- What is the reason for corrosion?
- Are cracks and structure deterioration on the increase?
- What is the expectation of the extent of deterioration of concrete and extension along the corrosion in the steel?
- What is the impact of current and expected future deterioration on the safety of structure?

DOI: 10.1201/9781003407058-8

Deriving the answers to these questions will need structural assessment methods and different measurement methods clarified in Chapter 4. These will help in deciding the cause of corrosion and present and future deterioration of the structure.

It is worth noting that the process of repair and restoration of concrete structures from the assessment phase to the execution phase needs a high level of experience. When conducted by inexperienced engineers, even the simple evaluation stage may lead to the wrong choice of repair method, which can become a big problem. A special report by the RILEM Committee (1994) helps clarify the various repair strategies used in most structure assessment cases:

- Re-establishment of the deteriorated member.
- Comprehensive repair of the concrete member to regain its ability to withstand full loads.
- Repair of a particular portion of the concrete member, followed up continuously over periods of time.
- Strengthening structures by an alternative system to bear part of the loads

In most cases, the strategy of repair is either a comprehensive or a partial repair of the concrete member. These strategies are common in the rehabilitation of concrete and they depend on the structural system, external environmental factors, and the degree of structural degradation.

8.2 MAIN STEPS TO EXECUTING REPAIR

In the case of repair projects, it must be ensured that the contract is a great deal more flexible than most normal construction contracts and expert supervision is constantly available to make decisions and to control and record the types and amount of work. Provisional items may be unavoidable, and premeasurement and "cost plus" work should be considered acceptable for this type of contract. It is as likely in corrosion repair or any brown field project as it is in human surgery that estimates of the full nature and extent of the problems will change once the patient has been opened up.

Repairs in some cases have to be carried out on structures that are in use. In the case of buildings, the people who occupy them will be trying to work, learn, sleep, or recover from sickness while the work is going on.

A little trouble taken to consult and inform the occupants about what is going to happen to their environment and their life during the repair process.

In the case of industrial plants, which are operated continuously, repairs can sometimes be divided into operations that can be done while the plant continues to operate and those that will have to wait for a shutdown. The same applies to airfield, highway, or railway sites, where working at night while traffic is light may be an additional option.

There are several regular steps in the repair of structures exposed to corrosion. The very critical first step is to strengthen the structure by performing structural analysis and designing a suitable location for the temporary support. The second step is to remove the cracked and delaminated concrete. It is important to clean the concrete surface and also the steel bars by removing rust. After rust is removed by brush or sand blasting, the steel bars should be painted with epoxy coating or replaced; then new concrete can be poured. The final step is to paint the concrete member for external protection. This is a brief description of the repair process. These steps are explained in detail in the following sections.

8.2.1 Structure Strengthening

One of the most dangerous and important first steps necessary for the repair is selecting the temporary support, which depends on the following:

- Evaluating the state of the whole structure.
- Determining how to transfer loads in the building and its distribution.
- Determining the volume of repair that will be done.
- Determining the type of concrete member that will be repaired.

As mentioned earlier, the repair process must be carried out by a structural engineer with a high degree of experience, one who has the capability to perform structure analysis and has powerful knowledge of the load distribution in the structure, according to the kind of repair. The structural engineer has the responsibility of choosing the right way to optimize the process of crushing and of determining the ability of the structure members to carry the loads that will be transferred to them. Therefore, the responsible engineer should design the temporary support based on data collected and previous analyses and should be cautious in the phase of execution of temporary supports.

Choosing how to remove the defective parts will be based on the nature of the concrete member in the building as a whole; any member of breaking concrete has a detrimental impact on neighboring members because the process of breaking will produce a high level of vibration. The temporary members must be strong and designed to withstand loads and must be transported easily and safely to the defective area. The entire structure depends on the design and execution of the temporary supports and their ability to bear loads safely.

8.2.2 Removing Concrete Cracks

There are several ways to remove the part of the concrete that has cracks on its surface and shows the effects of steel corrosion. These methods of removing the delaminated concrete depend on the ability of the contractor, the specifications, the cost of breaking, and the whole state of the structure. The selection of the breaker methods is based on the cause of corrosion; if it is due to carbonation or chlorides, then one must also consider whether cathodic protection should be performed in the future. In this situation, the breaking work would take place on the falling concrete cover; it would be cleaned and all the delaminated concrete and cracked concrete parts removed. Then, high-strength, nonshrinking mortar would be poured.

If the corrosion in steel reinforcement is a result of chloride propagation into concrete, most specifications recommend removing about 25 mm behind the steel and making sure that the concrete on the steel has no traces of chlorides after the repair process. The difference between good and bad repair procedures is obvious in Figure 8.1.

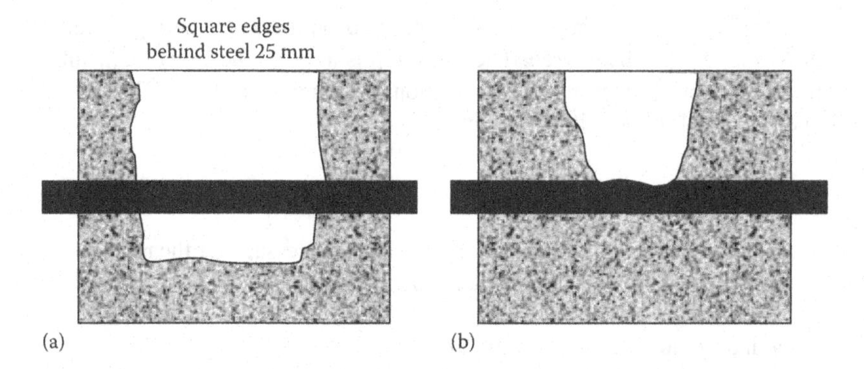

FIGURE 8.1 Differences between (a) good and (b) bad repairs.

The difference in the procedure of breaking the delaminated concrete is due to the difference in the causes of corrosion. Therefore, a careful study to assess the state of the structure and the causes of corrosion is very important to achieve a high-quality structure after the repair process. The evaluation process is the same as illness diagnosis. If there are any mistakes in the diagnosis, the repair process will be useless, as well as a waste of time and money. It is necessary to define the work procedure and quantity of concrete that will be removed. This step is considered one of the fundamental factors for designing and installing wooden pillars of a building to be used during repair. Therefore, the work plan must be clear and accurate for all the engineers, foremen, and workers who participate in the repair process.

After competent staff are secured and the building assessment and design of supports and ties are completed, some information may not be available—for example, the construction procedure for the building or the workshop drawings or specifications followed when the building was constructed. Therefore, the risks are still high. The only factor that can help in reducing this risk in spite of steel corrosion is the increase in concrete strength with time. However, this compensation occurs within a limit because the steel is carrying most of the stress and the risk will be very high in case of spalling in the concrete cover due to reduction in concrete cross-sectional dimensions.

It is necessary and important to remove concrete for a distance greater than the volume required for the removal of defective concrete so that proper steel can be reached. This will be important later in the repair process. Several methods are commonly used for breaking and removing the defective concrete, and these are explained next.

8.2.2.1 Manual Method

One of the simplest and easiest methods is to use a hammer and chisel to remove defective concrete. This is considered one of the most inexpensive ways, but it is too slow compared to mechanical methods. However, mechanical methods produce high noise and vibration, have special requirements, and need trained labor. Using the manual method makes it difficult to spare concrete behind the steel. The method is used in the case of small spaces and is preferred in the event of corrosion due to carbonation and attacking chlorides from outside, when it is not necessary to break the concrete behind the steel. Any worker can manually break the concrete, but it is necessary to choose workers who have done repair work before as they must be sensitive in breaking the concrete to avoid causing cracks to adjacent concrete members.

8.2.2.2 Pneumatic Hammer Methods

These hammers work using compressed air; they weigh between 10 and 45 kg. If they are used on the roof or walls, their weight will be about 20 kg. They need an attached small power unit to do the job, but in large areas may require a separate, bigger air compressor, as shown in Figure 8.2. This machine requires proper training for the

FIGURE 8.2 Air compressor.

FIGURE 8.3 Using a pneumatic hammer to remove wall concrete cover.

worker that uses it. Compressed air hammers have a few initial costs. A few have been discussed by the Strategic Highway Research Program (SHRP) and, based on Vorster et al. (1992), a research program of highways, the terms of the contract are governed by the contractor responsible for handling the breaking work.

The use of pneumatic hammers is more economical when a small, rather than large, area is to be removed (see Figure 8.3). A water gun is preferable for large areas (described in the following section). When the client does not specify the area that needs to be removed, the cost calculation will be based on square meters. In this case, the risk will be low because the machine's initial cost is small, so a pneumatic hammer would be preferred. Some performance rates are about 0.025–0.25 m^3/h using hammers weighing 10–45 kg, respectively. In summary, pneumatic hammers are useful when small areas of concrete are to be removed.

8.2.3 Water Jet

This method has been commonly used since it was introduced to the market in the 1970s. It relies on the existence of water at the worksite and on the removal of a suitable depth of concrete in a large area. It removes fragmented concrete, cleans steel bars, and removes part of the concrete behind the steel bars, as shown in Figures 8.4–8.6. The water jet is used manually by an experienced worker who has previously

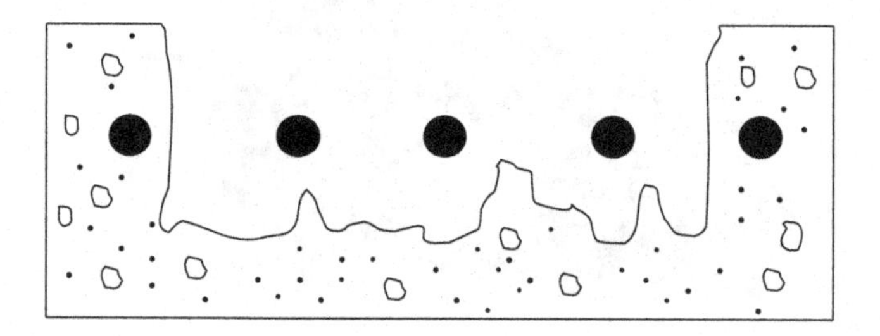

FIGURE 8.4 Shape of delaminated concrete when a water jet is used.

FIGURE 8.5 Concrete surface after using a water jet.

dealt with a hose, which is pushing water under high pressure or perhaps through a mechanical arm. Very high safety precautions need to be applied to the worker who uses it and the site around it.

The water used must not have any materials that can affect the concrete, such as high chloride ions; in general, it must be potable water. The water gun consists of diesel engines and pressure pump and connected to a hose that bears high water pressure, from 300 to 700 kg/cm^2 at the nozzle. At least 400 kg/cm^2 is required to cut the concrete, and the rate of water consumption is about 50 L/min. The performance rate of a water jet to break concrete is about 0.25 m^3/h—in the case of the use of a small pump—and can reach up to about 1 m^3/h if two pumps or one big pump is used.

8.2.4 Grinding Machine

This is used to remove concrete cover in the case of large, flat surfaces. An example is a bridge deck, as in Figure 8.7, but it must be done cautiously so that the process of breaking does not reach the steel. As in the case of any contact between the grinding machine and steel reinforcement, it will cut the steel bars and damage the machine. This method of breaking delaminated concrete is seldom used in the United Kingdom, because insulation film is often used to prevent water from

FIGURE 8.6 Surface after using a water jet.

FIGURE 8.7 Grinding machine.

seeping through, and it is used sparingly on U.S. bridges. The grinding machine is usually used after the water gun or the pneumatic hammer to obtain final concrete breakdown around and under the steel reinforcement. Therefore, one must take into account whether the thickness of the concrete cover is equal. The rate of removal of the concrete by this machine is very fast; it removes about 1 m^3/min and its cutting part is 2 m in width.

8.3 CLEAN CONCRETE SURFACES AND STEEL REINFORCEMENTS

This phase removes any remaining broken concrete with a process of cleaning. At the same time, the process of assessing the steel and cleaning up and removing corrosion from the roof takes place.

8.3.1 Concrete

The stage of preparing a surface by pouring the new concrete is one of the most important stages of the repair process. Before application of the primer coating, which provides the bond between the existing old concrete and the new concrete for repair, the concrete surface must be well prepared, and this takes place according to the materials used. In all cases, the concrete surface must be clean and not contain any oils, broken concrete, soil, or lubricants. If any of these elements are present, the surface must be cleaned completely through sand blasting, water, or manually using brushes. This stage is very important and very necessary, regardless of the type of material used to bond the new concrete with the old. If you neglect the preparation of the surface might affect the repair process as a whole because this stage of repair is the least expensive stage in the repair process.

It is not necessary to prepare the surface when a water gun is used to remove the delaminated concrete because the surface will be wet enough and will be clean after the crushing process. One of the benefits of using a water gun is that the air pressure cleans the surface of concrete and removes any growths on the fragmented concrete, this will create the surface material ready to adhere with the new concrete material.

When cement mortar or concrete is used in the repair, the concrete surface will be sprayed with water until saturation is reached, which usually takes 24 hours. This can also be done by wetting burlap. The water spray must be stopped and the burlap removed for about 1–2 hours, depending on weather conditions, until the surface is dry. It is then coated by a mix of water and cement only, which is called slurry and is applied by brush. Epoxy coating can also be used as an adhesive between new and old concrete; it is necessary to follow the manufacturer's specifications and warnings.

The preparation of the concrete surface to achieve a better bond between the old and new concrete is stated in the American specification ACI (American Concrete Institute) 503.2-79. This specification says that the preparation of surfaces to receive epoxy compound applications must be given careful attention as the bonding capability of a properly selected epoxy for a given application is primarily dependent on proper surface preparation. Concrete surfaces to which epoxies are to be applied must be newly exposed, clean concrete free of loose and unsound materials. All surfaces must be meticulously cleaned, be as dry as possible, and be at proper surface temperature at the time of epoxy application.

ASTM C881-78 is specific about the epoxy material that performs the bond; it must be well defined by supplier specifications identical to the different circumstances surrounding the project, particularly when, with an increase in temperature, change might occur in the resin, as well as the nature of loads carried by the member requiring repair. It is worth noting that during the execution of the repair process, the use of epoxies will be significant. Therefore, we must take into account the safety factor for workers who use such material. These workers must wear their personal protective equipment (PPE), such as gloves and special glasses, for safety issues because these materials are very harmful to the skin; epoxy can cause many dangerous diseases if one is exposed to it for a long time. Also, some epoxies are flammable with temperature, and that fact must be taken into account during storage and operation.

8.3.2 Clean Steel Reinforcement Bars

After removal of the concrete covers and cleaning the surface, the next step is to evaluate the steel reinforcement by measuring steel diameter. If the cross-sectional area of the steel bars is found to have a reduction equal to or more than 20%, additional reinforcing steel bars must be added. Before pouring new concrete, one must be sure that the development length between the new bars and the old steel bars is enough, as shown in Figure 8.8. It is usually preferable to link the steel by drilling new holes in the concrete and connecting the additional steel on concrete by putting the steel bars in the drilled hole filled with epoxy. However, in most cases, the steel bars are completely corroded and need to be replaced.

In the case of beams and slabs that require additional steel reinforcement bars, it is preferable to connect the steel bars with concrete by drilling new holes in the concrete and making the bond of the steel bars in the holes by using adhesive epoxies. For beam repairs, the additional steel bars are fixed in a column that supports this beam. In the case of slabs, the steel bars are fixed in the sides of the beam that is supporting the

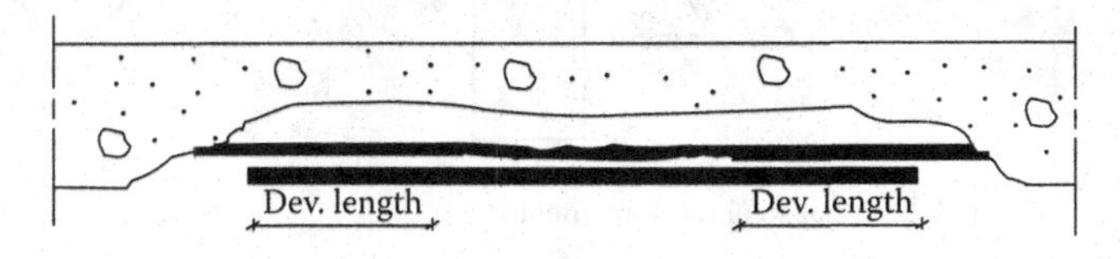

FIGURE 8.8 Installing additional steel.

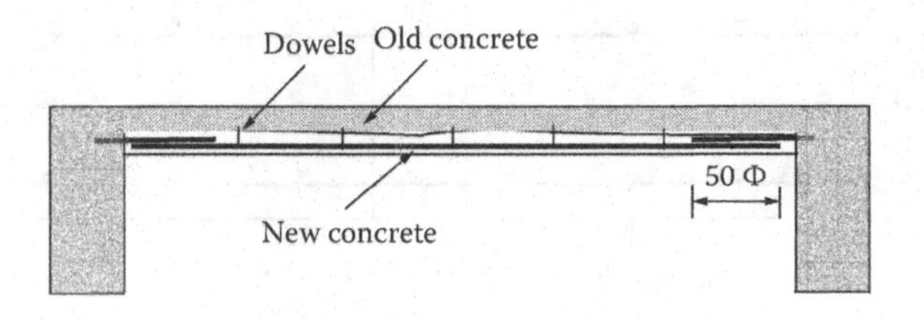

FIGURE 8.9 Steel reinforcement installation for slab repair.

slab, as shown in Figure 8.9. The dowelling will be fixed on the beam side by drilling holes to a depth of around 70–80mm; the dowel will be fixed in the holes by epoxy.

The depth of the hole may differ, depending on the type of the epoxy and the supplier's assumptions and recommendations. The hole's diameter should be 40mm greater than the bar's diameter to ensure that the steel bars are fixed completely. It is always 12 or 16mm; it is preferable to increase the bars' diameter and decrease their number to reduce the number of holes in the beams. The dowel is put in the drilled hole, as shown in Figure 8.9, with a length 50 times the bar's diameter, which overlaps the new steel bars and is fixed between them by the steel wire or welding. The bars are fixed at both directions by putting a small dowel from the slab in the intersection between the steel bars; these small dowels are also fixed to the slab by epoxy.

Most coastal cities have special creations by architectural firms, and design is based on placing all the balconies facing the coast for sea view and to increase the value of the units in the building. At the same time, balconies require distinctive structural elements in their design and execution. From a structural point of view, a balcony is cantilevered; the cantilever has the lowest redundancy of all structural elements, so it needs special attention in the case of repair and restoration.

In practice, balconies often have corrosion on the upper steel reinforcement, as it is the main steel, particularly in the cantilever. The tiles will be removed, the concrete cover removed, and then the reduction on the upper steel reinforcement will be inspected, as we mentioned earlier. If it is found that the steel bars are badely corroded, in this case it is mandatory to add a new upper steel reinforcements. These new steel bars must be extended on the adjacent bay bu a length equal to 1.5 times the cantilver length as shown in Figure 8.10. The dowels are installed vertically, with a distance of about 350mm

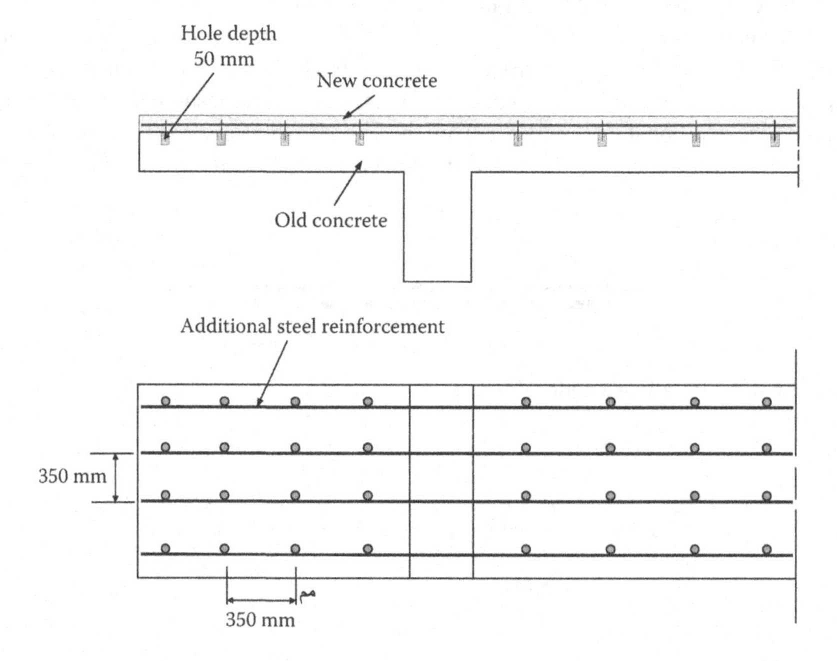

FIGURE 8.10 Cantilever repair.

between them. They are fixed by the concrete slab with epoxy to a depth of 50mm, and the new main steel bars are attached to it.

It is worth mentioning that in all repair processes—and especially in cantilevers because they are a point of weakness in terms of structure safety—one must pay attention to the design and execution of the wood form and temporary supports, as well as the part of the building that will be overloaded during the repair process. It is necessary to consider the structural system and stress distribution in different elements, the weak points in the building, the building's maturity, and the method of repair that is applied.

The work of repairing beams is almost the same as a slab repair process. The difference is in fixing the dowels in the columns, as shown in Figure 8.11, and fixing the stirrups by making holes with a depth of around 50–70mm (according to the epoxy manufacturer's recommendation). Epoxy is applied on both sides of the beam to stabilize stirrups, about 8 mm/200mm, as shown in Figure 8.11. After that, as a normal procedure, epoxy coating is applied as an adhesion between old and new concrete and then the new concrete is poured manually, or by shotcreting. As shown in the figure, the beam is strengthened by increasing the cross-sectional dimension and adding steel reinforcement.

As for the columns, the repair is performed as shown later in Figure 8.14, which shows the casting on-site; however, the minimum distance allowed to cast concrete easily is about 100–120mm from each side. The dowel is fixed for the first floor, as in Figure 8.12; this does not require drilling holes in the base, but rather resting the legs of the steel bars on the foundation. The steel bars will be distributed around the column circumference, as in the figure, to cope with the reduction of steel reinforcement

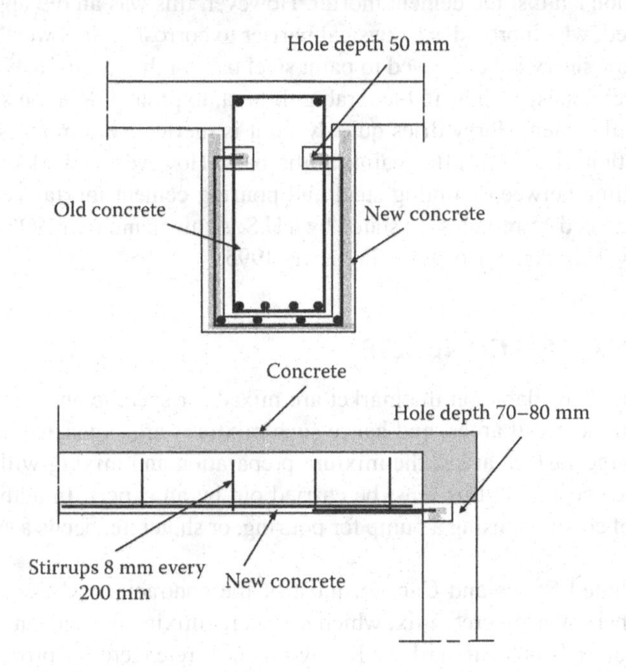

FIGURE 8.11 Beam repair.

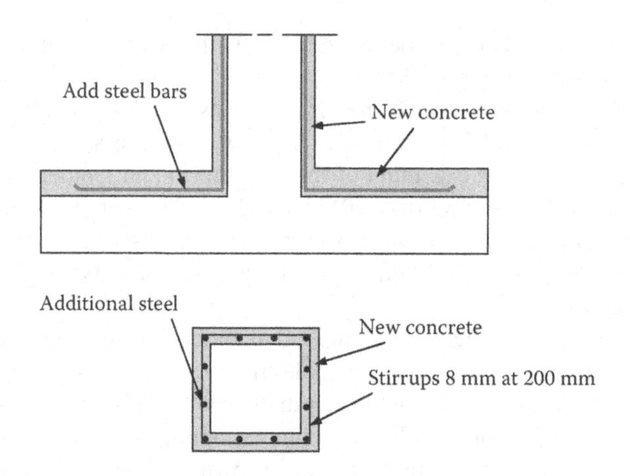

FIGURE 8.12 Repairing a concrete column.

cross-sectional area due to corrosion. The steel cross-sectional area percentage to the concrete column cross-sectional area is the same as that in the old column member and not less than as stated in the code.

Note that when a water jet is used, it will clean and remove corrosion from the steel; if it is not used, sand blasting can be used to remove corrosion. Then, the old steel can be painted by using epoxies after cleaning the steel bars completely, especially from the effects of chlorides. Another way to protect steel reinforcement bars is by using slurry, which is a mixture of rich cement, water, and paint, for alkaline protection against the cement mortar. However, this was an old approach. Now, epoxy is used, which provides a physical barrier to corrosion. It is worth mentioning that until now slurry is being used to paint steel bars in the case of new construction in very hot climates, such as in the Arab gulf area, to protect them on site. Note that the improved cement slurry dries quickly, so it is ineffective in repairs that require the installation of a form after painting the bars. However, it works well in cases where the time between painting steel and pouring cement mortar is short, which should not exceed 15 minutes, as stated by a U.S. Army manual in 1995 (Department of the Army, U.S. Army Corps of Engineers 1995).

8.4 NEW PATCH CONCRETE

Some mixtures available on the market are mixed for specific applications for easy use in repairs to small areas, and hence such mixtures are expensive. In the case of repairs to large surface areas, the mixture preparation and mixing will be done on-site to reduce costs, but this must be carried out by an expert. In addition to these properties of concrete, using a pump for pouring, or shotcrete, needs a special design mix.

In the United States and Canada, most of the contractors who work on bridge repair use their own concrete mix, which has secret mixing proportions based on the available materials in local markets. Ready-mix concrete factories provide guarantee in the case of corrosion due to carbonation, but not when it is due to chlorides, for fear of the presence of chloride after the repair process is complete. Most manufacturers of materials used in mixing are field execution contractors, but only when they supply materials will they provide all the information and technical recommendations for the execution, performance rates to calculate the required amount of the materials, and appropriate method of operation.

Worldwide, all companies operating in the construction field and dealing with chemicals must have a competent technical staff that can assist on-site personnel or supervise workers to avoid any errors and to define responsibility if defects are found. Two types of materials are used as a new mortar for repair: polymer mortar or cement mortar enhanced by polymers. Both are described in the following sections.

8.4.1 Polymer Mortar

This mortar provides the specific components required for repair, ease of operation, and control over setting time. It binds well with the existing old concrete and does not need any other element for this cohesion. This mortar has a high compression strength of around 50–100 MPa and a high tensile strength as well.

Despite the many advantages of such mortar, their properties are different from those of existing concrete as they have a thermal expansion coefficient equal to 65×10^{-6}°C, corresponding to a concrete thermal expansion coefficient equal to 12×10^{-6}°C. Moreover, the modulus of elasticity for mortar is much less than in the case of concrete. These differences in the properties lead to the presence of cracks as a result of internal stresses; however, they can be overcome by aggregate sieve grading to reduce the proportion of polymers, which reduces the difference in their natural properties. There are many tests to define this mortar and its noncompliance with the concrete (e.g., ASTM C 884-92: "Standard Test Method for PCBs: Comparability between Concrete and Epoxy-Resin Overlay").

8.4.2 Cement Mortar

Polymer mortar provides physical protection to steel reinforcements; however, cement mortar provides passive protection as it increases the alkalinity around the reinforcement. The trend now is to make optimum use of cement mortar for repair as a result of corrosion because it will have the same properties as the existing concrete. In addition, its passive protection of steel works against corrosion.

Some polymers are added to the cement mortar in liquid or powder form to improve its properties (ACI Committee 548 1994; Ohama 1995) and to increase flexural resistance and elongation, reduce water permeability, increase the bond between old and new concrete, and increase the effectiveness of its operation. The polymer used is identical to cement mortar mixture; the properties that control the polymers in the cement mortar are stated in the ASTM C 1059-91 specification for latex agents for bonding fresh to hardened concrete. Silica fume can be used with the mixture, as well as superplasticizers, to improve the properties of the used mortar and to reduce shrinkage.

8.5 EXECUTION METHODS

There are several ways to implement the repair process, which is entirely dependent on the type of structure member to be repaired and the materials used in the repair. These methods are described in the following sections.

8.5.1 Manual Method

The manual method is used in most cases, especially when small spaces are to be repaired (Figure 8.13). A wooden form can be made and then concrete can be poured into the damaged part, as shown in Figure 8.14, in the case of concrete columns or vertical walls. One must take into account the appropriate distance to cast the concrete in the wooden form easily. This method is commonly used for its efficiency and low cost. It does not require expensive equipment, but extensive experience in fabrication, installing the wooden form, and the casting procedure is needed.

8.5.2 Grouted Preplaced Aggregate

As shown in Figure 8.15, the aggregate should be placed with gap grading in the area to be repaired. The next step is to make a grouting fluid by pumping it inside the aggregate through a pipe to fill the gap between the aggregate and the grout. This method of repair is used in repairing bridge supports and other special applications. This method requires special equipment such as pipe injection, pump, and other special miscellaneous equipment. Therefore, one can conclude that this method is used by private companies with high potential.

FIGURE 8.13 Manual method for repair.

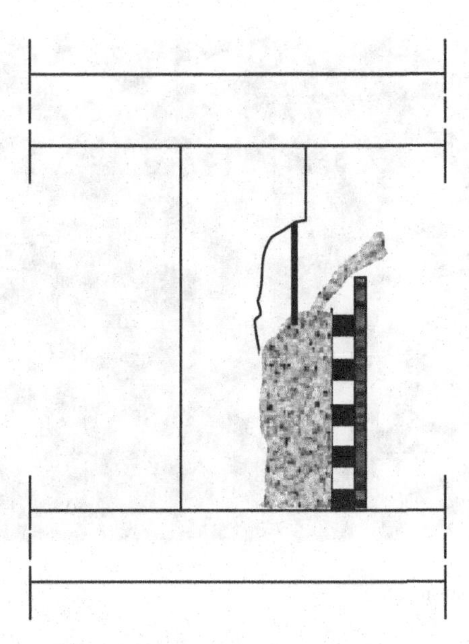

FIGURE 8.14 Casting concrete on-site.

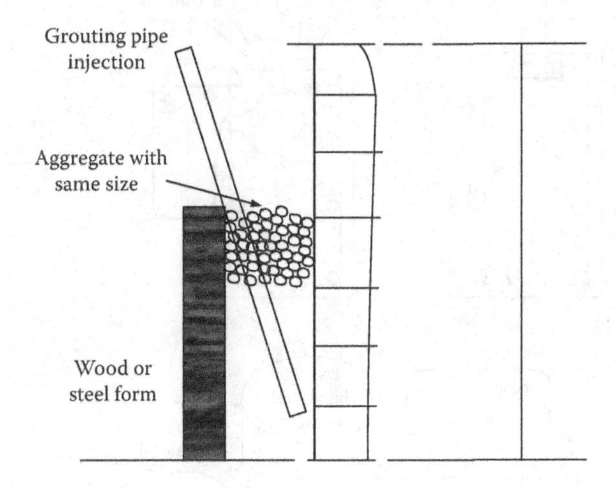

FIGURE 8.15 Injected preplaced aggregates.

8.5.3 Shotcrete

Shotcrete is used when large surfaces need to be repaired, but the mix and components of concrete suitable for use in shotcrete need special additives and specifications. As shown in Figure 8.16, health and safety precautions must be carefully followed for the worker using shotcrete, as it contains polymers and special additives. In the concrete mix design, the nominal maximum coarse aggregate size must be defined to suit the shotcrete equipment's nozzle and pump to avoid any problem during concrete casting. Complete member casting is used when total reconstruction is

FIGURE 8.16 Using shotcrete.

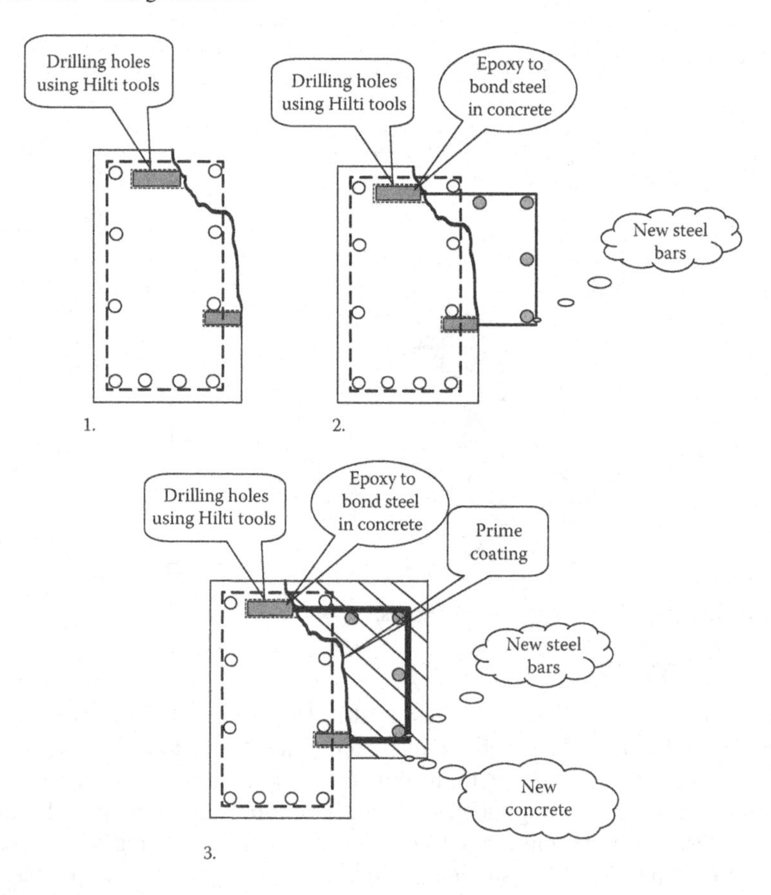

FIGURE 8.17 Steps of repair.

required for the concrete member whose steel reinforcement bars have been depleted due to corrosion. Therefore, it is necessary to pour concrete for the complete member with full depth, as shown in Figure 8.17.

Frequently, in practice (e.g., when bathrooms are repaired), the concrete slab will be in a very bad condition, so an ordinary repair procedure will not be efficient. In this case, the entire concrete slab will be demolished, a new steel reinforcement will be installed, and the concrete slab will be poured, as shown in Figure 8.17. A decision to use this method must take into consideration the whole building's condition. This method of repair is usually applicable for bathroom slabs in residential buildings, because such slabs are usually designed as simple supports for 100 mm diameter plumbing pipes.

8.6 REPAIR STEPS

Figure 8.17 presents a case study for repairing a reinforced concrete ring beam supporting steel tanks containing oil, as in the case study in Chapter 4. The ring beam is structurally carrying a tensile strength due to an earth load affecting the beam, which puts the ring beam under tension. Corrosion of the steel reinforcement bars means reduction of the capacity of the ring beam. Thus, the best solution is to increase the steel bars to overcome the reduction on the steel cross-sectional area. The process of repair takes place in three steps. In step 1, holes are drilled and steel stirrups are put in the holes and fixed by epoxy. In step 2, the main steel bars are installed and the concrete surface is coated with epoxy so that the existing old concrete will bond with the new concrete. In step 3, the new concrete is poured with a proper mixing design, which is improved by additives such as polymers.

8.7 NEW METHODS FOR STRENGTHENING CONCRETE STRUCTURES

There are other ways to strengthen reinforced concrete structures, including traditional methods such as the use of steel sections. This method has many advantages, including the fact that the thickness of the concrete sections need not be increased significantly. This is important to maintain the building's architectural design. On the other hand, it is a quick solution to strengthen the concrete member and is usually used in industrial structures and buildings. If it is required to do strengthening due to corrosion in steel reinforcement, the solution of placing steel supports is not considered appropriate but can be used, taking into account that it requires epoxy paints periodically.

Recently, many studies have been conducted on using fiber-reinforced polymer (FRP), which has many advantages over steel. Most importantly, it does not corrode and therefore can be used in any environment exposed to corrosion. There are some modern ways to protect steel reinforcements from corrosion, such as replacing steel reinforcement bars with bars made from fiber polymer and also by coating the concrete surface with epoxy paints.

Generally, the goal of strengthening the structure is to restore the concrete member's ability to withstand loads and to carry loads greater than the loads for which it was designed or to reduce the deflection that occurs when the concrete member is overloaded. The strengthening can proceed due to the presence of cracks and needs to stop the increase of cracks or reduce crack width in the concrete member. Therefore, strengthening is required in the following cases:

- An error in the design or execution causes a reduction in the steel cross-sectional area or a decrease in concrete section dimensions.
- Deterioration in the state of a structure leads to weak resistance (an explicit example of this is the corrosion in steel reinforcement).
- An increase in loads increases the number of stories or changes the building's function (e.g., from residential building to office building).
- The structural system is changed (e.g., removing walls).
- Holes are made on the slab, reducing slab strength.
- Some special structures need strengthening, for example, bridges; the stress of the flow of traffic, which increases with time, and heavier trucks may be higher than what the bridge was designed for, so it needs to be strengthened to accommodate the overloaded and new operation mode.
- The load may change with time, such as a building in an earthquake zone. This happened in Egypt when the earthquake map was changed and, consequently, some buildings needed to be strengthened to match the new specifications.
- The structure may be exposed to high temperature due to fire.
- A big machine may need to be installed in an industrial structure.
- In industrial buildings, the machines mature over time or change their mode of operation, resulting in high vibration, so the structure needs to be strengthened.

8.8 USING STEEL SECTIONS

Using steel sections on a large scale to strengthen reinforced concrete structures has different advantages. The method of strengthening the concrete member varies depending on the member that needs to be strengthened. The strengthening process is calculated based on the reduction in strength in the concrete section, and this reduction is compensated by using steel sections.

If a concrete beam needs to be strengthened due to lack of steel reinforcement, a I-beam can be used (Figure 8.18) when a big change in architectural view is not warranted and there is a minor reduction in member capacity. On the other hand, in the case of a significant reduction in concrete member capacity and a need to increase the depth significantly, an I-beam section will be used (Figure 8.18). Another way to strengthen beams is by stabilizing sheets of steel (Figure 8.19) and fixing them using mechanical bolts or through chemical substances such as epoxy. The purpose of adding this steel sheet is to increase the moment of inertia of the beam and thereby increase its ability to withstand stress greater than the flexural stresses for which it was designed. In the case of beams that have a problem in shearing force direction, they can be strengthened (Figure 8.20).

Cracks present on the upper surface of the slab in the beam direction indicate a shortage in the upper reinforcement of the slab; it can be strengthened by using steel sheets on the upper surface, as shown in Figure 8.21. These plates are fixed by bolts

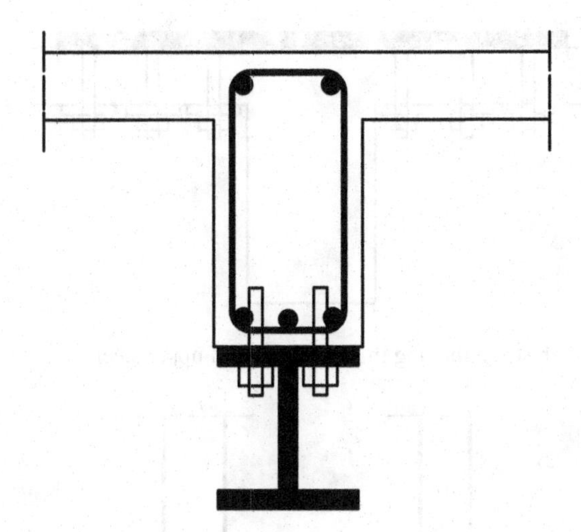

FIGURE 8.18 Beam strengthening.

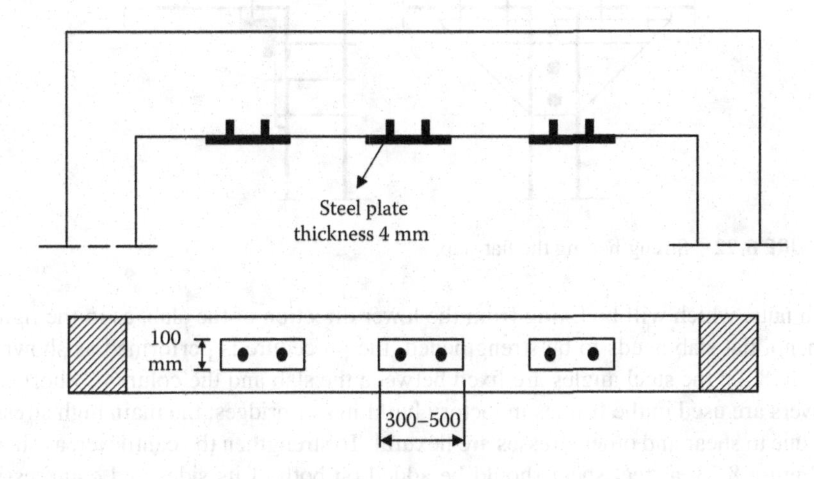

FIGURE 8.19 Strengthening a concrete beam with steel plate.

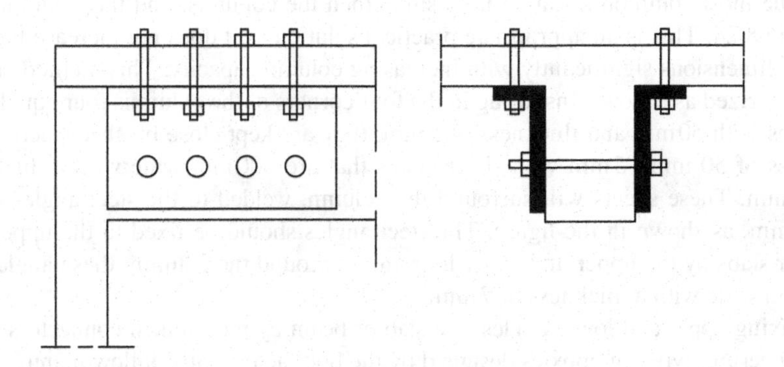

FIGURE 8.20 Beam strengthening in the direction of shear.

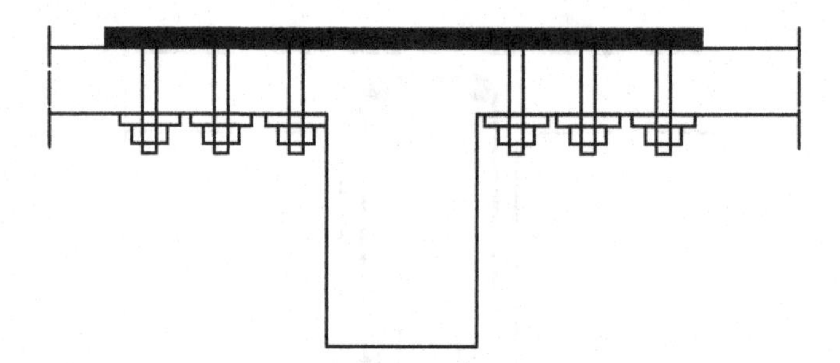

FIGURE 8.21 Slab strengthening in the direction of upper steel.

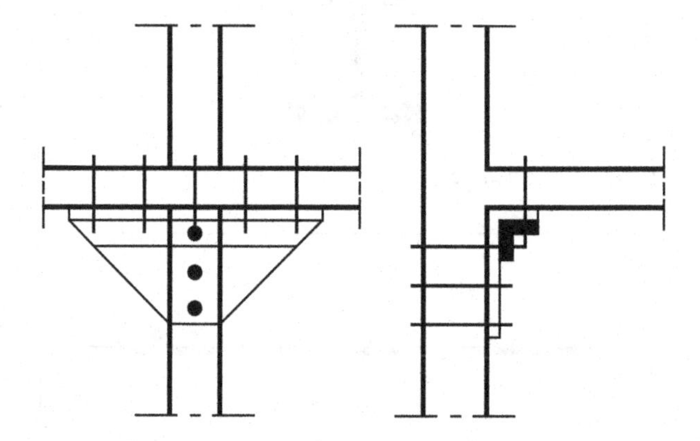

FIGURE 8.22 Strengthening the flat slab.

with nuts, which will be fixing from the lower direction of the slab, as in the figure. When a flat slab needs to be strengthened, the procedure is performed as shown in Figure 8.22; the steel angles are fixed between the slab and the columns. Short cantilevers are used in the frames in factory buildings or bridges; the main high stresses are due to shear and other stresses are flexural. To strengthen the cantilever, as shown in Figure 8.23, a steel sheet should be added on both of its sides and compression force applied with bolts, thereby reducing the likelihood of a collapse due to shear.

The most common situation is to strengthen the columns, and the solution is in Figure 8.24. This is an appropriate practical solution as it does not increase the column dimensions significantly with increasing column capacity. This method can be summarized as follows: Installing in the four corners of the columns four equal steel angles with 50 mm and thickness of 5 mm; they are kept close by sheets across the angles of 50 mm × 5 mm with sheet plates that are 500 mm wide with a thickness of 5 mm. These sheets will surround the column, welded to the steel angles every 200 mm, as shown in the figure. The steel angles should be fixed to the upper and lower slabs by the upper and lower four angles around the column; these angles are 70 mm wide with a thickness of 7 mm.

Fixing upper and lower angles to a slab or beam by mechanical connectors or by using certain types of epoxies designed by the bolt factory must follow manufacturer recommendations. It is important to know that these bolts transfer the load from up

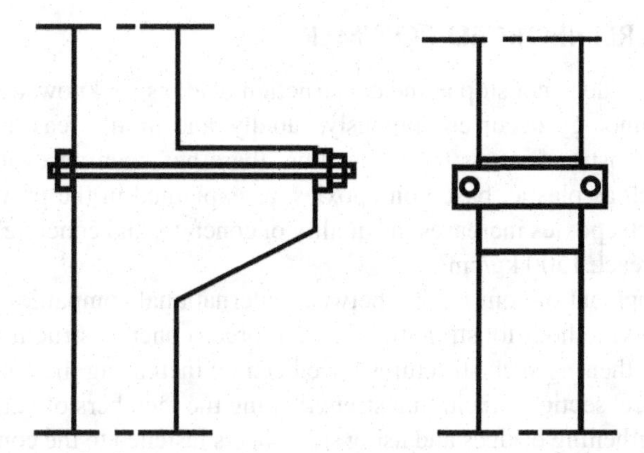

FIGURE 8.23 Strengthening short cantilevers.

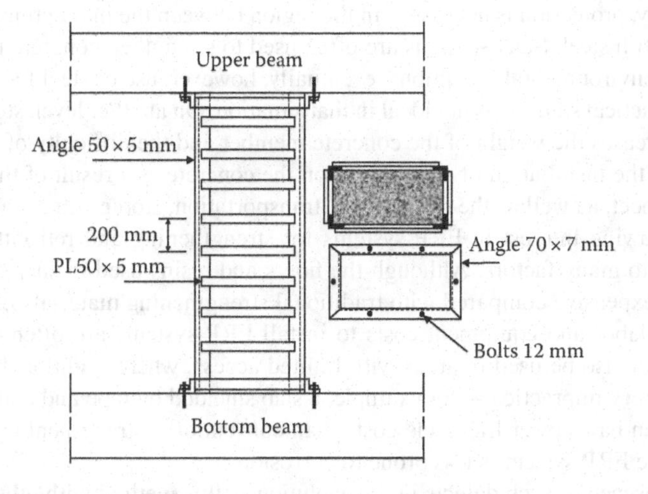

FIGURE 8.24 Strengthening a reinforced concrete column.

to down. In the strengthening methods discussed earlier, the dimensions of the steel selected are as small as possible so as to maintain the architectural view; these can be covered by using wire steel mesh and then applying plastering, covering the surface by wood, or using special plastering or decoration methods to provide an acceptable view.

In cases of slab deterioration due to corrosion, it can be strengthened by using a hot rolled steel section, which will be installed at the bottom of the concrete slab to carry it in a short direction and the distance between these steel beams should be minimized to reduce the steel beams sections. The I-beam can be used, but the use of a channel section is preferred, whose maximum depth, practically, should be 150 mm. For decoration, a false ceiling with lighting or sound systems can be used to obtain the most benefit. The steel beam is inserted into the reinforced concrete beam and using the epoxy for fixation and in some cases you can use mechanical or epoxy steel anchor. The gap between the steel section and the concrete slab will be filled with cement grout.

8.9 FIBER-REINFORCED POLYMER

Modern science does not stop at the construction field; as we know, developments in computer technology occur continuously, rapidly, and in all areas, including engineering and construction. At the same time, there has been an evolution in some materials, such as plastic, based on epoxies, as explained in the previous chapters. The use of such epoxies increases the quality of concrete, and concrete's compressive strength can reach 1500 kg/cm^2.

The development of competition between international companies led to the creation of various methods for strengthening reinforced concrete structures. Traditional ways of strengthening such structures have been by increasing the concrete sections or by using steel sections, including strengthening the members of various constructions or strengthening bridges and using steel sheets fastened to the concrete by bolts. The only disadvantage for using a steel section for concrete strengthening is the corrosion of the steel section. Therefore, it needs to be protected from corrosion; particularly, protection is necessary in the region between the interacting surfaces of concrete with steel. Steel sections are often used to strengthen concrete members in corrosive environmental conditions; eventually, however, use of steel for strengthening as a practical solution is not ideal in that situation. On another level, steel is heavy, which increases the weight of the concrete member and the difficulty of strengthening due to the installation of steel sheets on the concrete as a result of the weight of the steel sheet, as well as the difficulty of transportation, storage, and stabilization.

The growing interest in FRP systems for strengthening and retrofitting can be attributed to many factors. Although the fibers and resins used in such systems are relatively expensive compared with traditional strengthening materials like concrete and steel, labor and equipment costs to install FRP systems are often lower. FRP systems can also be used in areas with limited access, where traditional techniques would be very impractical—for example, a slab shielded by pipe and conduit. These systems can have lower life-cycle costs than conventional strengthening techniques because the FRP system is less prone to corrosion.

The presence of such defects in the evolution of this method with other materials has led to resorting to the use of plastic fiber bars. FRP can serve as an alternative to steel sheets and has a wide range of advantages in the strengthening process. Its distinctive characteristics include its resistance to corrosion, which occurs under any environmental circumstances, considering to avoid heat or electricity to reach the concrete member, and resistance to chemicals, which makes it a solution suitable for industrial structures. There are different types of FRPs; a popular type is carbon fiber-reinforced polymer (CFRP), which is most commonly used in practical applications because of its unique properties in terms of resistance and the resistance with time, as well as resistance to stress.

8.9.1 CFRP

CFRP has certain commercially accepted dimensions: commonly 1–1.5 mm, with 50–150 mm width. All of these types of polymers consist of 60%–70% carbon fiber on one side with 10 μm diameter embedded in epoxy resin. The mechanical properties of CFRP differ from one type to another, but generally, the modulus of elasticity has a value between 165 and 300 N/mm^2 and the tensile strength has a value range of 2800 N/mm^2 to a lower value of 1300 N/mm^2. We find that the lower value of the modulus of elasticity corresponds to the maximum tensile strength value.

The carbon fiber sheet shown in Figure 8.25a is embedded in epoxy resin; the other sheet type will be fixed directly by epoxy to the concrete surface as presented in Figure 8.25b. The important property is its density, which is about 1.6 t/m^3, ten times that of steel's density of 7.8 t/m^3. This is a big difference and it enables easy transfer and fixing. The adhesive material used in the sheets of carbon fiber resin is

(a)

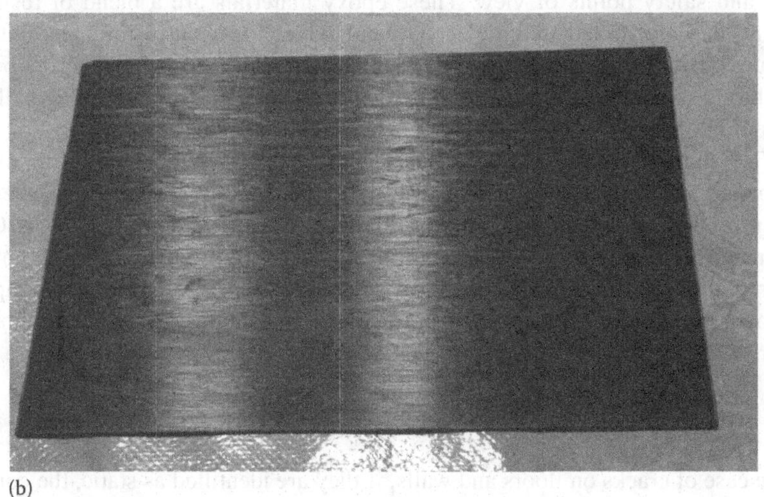

(b)

FIGURE 8.25 Carbon fiber sheet. (a) Fiber textile. (b) Fiber sheet.

a mixture of high resistance and filling quartz; its tensile strength is about 30 MPa, up from the concrete tensile strength by about ten times. Also, the rate of shrinkage and creep is small and can bear high temperatures and exposure to chemicals. The mechanical properties of carbon fiber sheets in the longitudinal direction are often controlled by the fiber, and its behavior is linear elastic until collapse.

8.9.2 Application On-Site

Before starting to work on strengthening a concrete structure, it is necessary to examine fully all the circumstances surrounding the structure in order to guarantee that strengthening the structural work is correct and that the company supplying the plastic fiber installs it utilize highly experienced personnel to design and identify segments of the required thickness. If strengthening is needed as a result of steel corrosion or due to chemical attack, it is important, for the success of the strengthening process, that rust from steel bars and chemical substances that attack the concrete be removed before carrying out the strengthening procedure.

The concrete surface must be settled through sand blasting to level the surface, eliminating any gaps or delaminated concrete. After sand blasting, the next step is to collect all the damaged concrete and sand by air suction, air pressure, or water pressure to clean the surface. Before fixing the CFRP, it is strongly recommended that the surface be level and clean; in the case of any cracks or delaminated concrete, the surface must be repaired before fixing the CFRP.

The mixture that provides the bond between the segments of CFRP and the concrete surface must be prepared. This mixture is very important in the repair as it is responsible for transferring the stresses from the concrete to CFRP. Therefore, the preparation of the epoxy mix and the percentage of mixture components must be precise and must be mixed under the supervision of the manufacturer because much experience in application, transportation, and storage is required from technical and safety points of view. These epoxy materials are a blend of resin and solvent, and are mixed by an electric hand mixer for about 3 minutes, until the mixture reaches homogeneity can be seen clearly, if the mixing is near the bottom and the sides of the container, it will be slower and generally greater speed allowed is about 500 rpm.

When the epoxy is applied to the surface, its thickness is about 0.5–2 mm; it must be applied accurately, making sure that all the gaps have been filled and there are no air voids. The images in Figures 8.26 and 8.27 show how to install the CFRP beam or a roof of reinforced concrete installed on a reinforced concrete column. The repair of the beam due to overshear stress or reduction of shear stress capacity is shown in Figure 8.26, and repair of the concrete column with CFRP is shown in Figure 8.27.

8.10 EPOXY FOR REPAIR

In the case of cracks on floors and walls, if they are identified as static, the concrete floor or concrete or masonry wall is injected with epoxy. In the case of floor, it will be injected continuously along the crack length, as shown in Figure 8.28, and in three

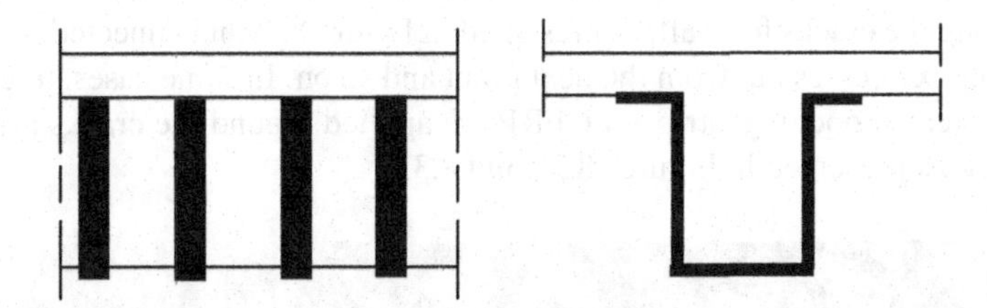

FIGURE 8.26 Strengthening beam for shear failure.

FIGURE 8.27 Column repair.

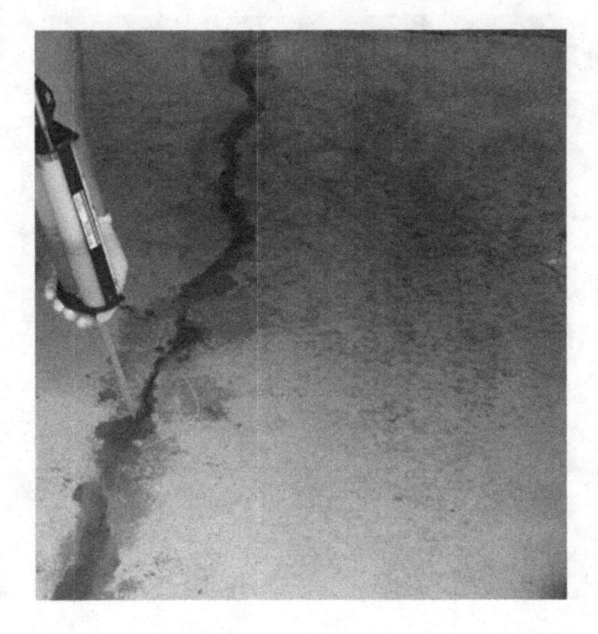

FIGURE 8.28 Repair of wall with epoxy.

points along the cracks for wall, as presented in Figure 8.29; it is injected at one point until the epoxy comes out from the next point and so on. In some cases, to guarantee there is no crack opening, strips of CFRP are applied around the cracks after epoxy is injected, as presented in Figures 8.30 and 8.31.

FIGURE 8.29 Repair of floor with epoxy.

FIGURE 8.30 Repair of wall cracks by CFRP method 1.

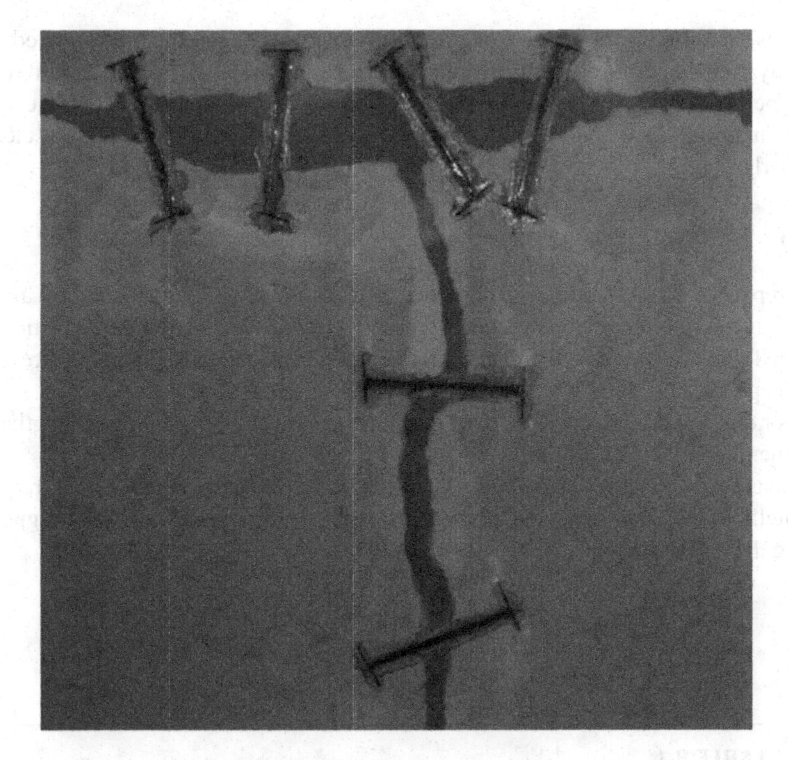

FIGURE 8.31 Repair of wall cracks by CFRP method 2.

8.11 GENERAL PRECAUTIONS

Note that the safety of a structure relies primarily on the quality of the repair process in terms of planning, design, and execution. As mentioned earlier, the repair process, identifying the necessary temporary support of the building and the location, depends on the nature of the structure and its structural system. Also, determining the method of breaking the defective concrete and removing it, as well as choosing the materials that will be used in the repair, must be in full conformity with the state of the structure and nature of the site of the structure, or location of the retrofit members within the overall structural system. All these elements require special expertise; otherwise, the possibility of risk in terms of structural safety is very high. Also, an ineffective repair process can have serious economic impact.

From the economic point of view, it must be remembered that the repair processes in general are high in cost in terms of materials used and the precautions that must be taken during execution, or in terms of trained workers who must carry out the repair process. Building safety and the health and safety of the workers are crucial; often epoxy materials or polymers can cause breathing and skin problems with time. Unfortunately, we find that health and safety rules in developing countries do not receive the same degree of attention as in developed countries—despite the fact that this results in loss of time and money for the whole project and represents a great danger to the management of the project.

It is also important not to forget that all equipment used must be cleaned thoroughly. Leaving equipment uncleaned after finishing the work will be a main reason for operational problems in the future. This is due to using epoxies, polymers, and other materials that can become problematic for machines if they are not cleaned properly.

8.12 REPAIR DESIGN

For repair or strengthening the concrete member needs to do a calculation and the following strength reduction factor is presented in ACI 562-19 and the reduction factor in Table 8.1 which is different than the assessment reduction factor as presented in Chapter 4.

For the load factors it will be used the same in case of assessment or strengthening the member which are as follow:

In case of external reinforcing systems susceptible to damage, the required strength of the structure without such external reinforcement shall be designed by using these load factors.

$$\phi Rn \geq 1.1D + 0.5L + 0.2S$$

$$\phi Rn \geq 1.1D + 0.75L$$

TABLE 8.1
Max. Strength Reduction Factor for Assessment

Strength	Classification	Transverse Reinforcement	f
Flexural, axial or both	Tension controlled[a]		0.90
	Compression controlled[b]	spiral	0.75
		other	065
Sheaf, torsion, or both			0.75
Interface shear			0.75
Bearing on concrete[c]			0.65
Post-tensioned anchorage zones			0.85
Struts, ties, nodal, zone, and bearing areas in strut-and ties model			0.80

[a] Applies when the steel tensile strain at member failure exceeds 2.5 ε_y where, ε_y is the yield strain of the tensile reinforcement.

[b] Applies when the steel tensile strain at member failure does not exceed ε_y for sections in which the net tensile strain in the extreme tension steel at nominal strength is between the limits of compression-controlled and tension-controlled sections, linear interpolation of φ shall be permitted.

[c] Does not apply to post-tensioned anchorage zones or elements of structs and tie model.

where D, L, and S are the effects due to the dead, live, and snow loads, respectively, and Rn is the nominal strength of the structural member computed using the material properties without the contribution of the external reinforcing system.

To be considered in case of live load has a high likelihood of being a sustained load, which mean take a time period more than 2 weeks in this case, as shown in the following equation:

$$\phi Rn \geq 1.1D + 1.0L + 0.2S$$

$$\phi Rn \geq 1.1D + 1.0L$$

In case of structural members with external reinforcement, the loads factor will be as follows:

$$\phi exR \geq (0.9 \text{ or } 1.2)D + 0.5L + 0.2S \qquad (8.1)$$

where $\varphi ex = 1.0$; R is the nominal resistance of the structural member, computed using the probable material properties during the fire event and considering the contribution of external reinforcement and S is the specified snow load. The dead load factor of 0.9 shall be applied when the dead load effect counteracts the total load effect.

The additional live loads incurred during a fire shall be considered, with a load factor of 1.0.

Internal forces and imposed deformations due to thermal expansion during the fire event shall be considered, with a load factor of 1.0, in determining the capacity on the structural system.

REFERENCES

ACI Committee 548. 1994. State of the art report on polymer modified concrete. In *ACI Manual of Concrete Practice*. Detroit, MI: American Concrete Institute.

Department of the Army, U.S. Army Corps of Engineers. 1995. *Engineering and Repair of Concrete Structures*, DC Manual No. 1110–2–2002. Washington, DC: U.S. Government Printing Office.

Ohama, Y. 1995. *Handbook of Polymer-Modified Concrete and Mortars*. Berkshire, England: Noyes Publications.

RILEM Committee. 1994. 124-SRC: Draft recommendation for repair strategies for concrete structures damaged by steel corrosion. *Materials and Structures 27(171)*:415–436.

Vorster, M., J. P. Merrigan, R. W. Lewis, and R. E. Weyers. 1992. *Techniques for Concrete Removal and Bar Cleaning on Bridge Rehabilitation Projects, SHRP-S-336*. Washington, DC: National Research Council.

9 Concrete Structure Integrity Management

9.1 INTRODUCTION

In general, the economic factor is one of the most influential in an engineering project. The economic cost of repairing a concrete structure as well as selecting a suitable system to protect the steel reinforcement from corrosion is one of the major factors influencing the choice among various alternatives for repair or protection. Virtually every method has an expected lifetime. By knowing the structure's lifetime, one can easily calculate the number of maintenance projects expected to be carried out throughout that time. For a new structure, it is important to conduct an economic study to choose among the alternatives for structure protection by considering the lifetime of the protection method with respect to the structure's lifetime. In addition, it is necessary to take into account the initial costs of periodic maintenance.

Generally, the cost calculation is based on a summation of the initial costs of protection and the cost of maintenance that will be performed in different periods and periodically. The number of maintenance times during the lifetime of a structure varies, depending on the method of protection and the method of repair procedure during maintenance. On the other hand, when one wants to perform maintenance to restore an existing old structure to its original strength, the method of rehabilitation is governed by the initial costs of repair in addition to the number of times the maintenance will be done and its costs over the remaining lifetime of the structure.

The previous chapters discussed how to choose among the various alternatives for protection methods as well as the appropriate repair methods and the materials usually used in repairs. The previous discussion was from a technical point of view. However, this chapter will discuss the way of comparing between different alternatives from an economic point of view to assist in the decision-making procedures. Therefore, an appropriate method will be used to compare alternatives to clarify the economic factors that affect the method of calculation.

As we have stated before, the lifetime of maintenance or protection of a structure must be taken into consideration, so the required time to perform maintenance will be discussed clearly as the first principle to calculate this lifetime. A practical example illustrating the comparison between different alternatives will also be applied. In this example, differences between the methods of protection for different structural economies will be detailed.

The process of determining the time of maintenance depends on the maintenance cost estimate versus the probability of structure failure; therefore, it is important to decide on the selection of an appropriate time to perform maintenance. The method of decision-making will be clarified as well as of using it from the standpoint of

DOI: 10.1201/9781003407058-9

determining the right time for maintenance in a less expensive verification process called 'optimization procedure.'

9.2 BASIC RULES OF COST CALCULATION

There are several basic rules for calculating the economic cost of an engineering project as well as for selecting the type of protection and method of concrete structure repair. The most popular methods of economic analysis are the present value, future value, and interest rate of return. In this section, the method of calculating the present value is described briefly, as it is the easiest way to select the appropriate method of economic repair and an appropriate system to protect the structure from corrosion.

9.2.1 Present Value Method

The cost of protecting reinforced concrete from corrosion consists of the preliminary costs of the method of protection and the money paid at the beginning of construction. The cost of maintenance and repair, on the other hand, will be incurred over the lifetime of the structure. In many cases, the cumulative cost of maintenance and repair is higher than the initial costs. In many projects, cost calculation is often based on the initial costs only, but it results in a total cost that is very high compared to the structure's cost estimate, which is only the initial cost.

The method of present value is used to calculate the present value of future repair, including the cost of equivalent current value, with the assumption that the repair will take place after n number of years:

$$\text{Present value} = \text{Repair cost}\,(1+m)^{-n} \tag{9.1}$$

where m is the discount rate, which is the interest rate before inflation. For example, assuming that the interest rate is 10% and the inflation rate is 6%, the discount rate m is equal to 4%, or 0.04.

The entire structural cost consists of the initial cost, which is called capital cost (CAPEX), and the sum of the present values of future costs due to maintenance, which is called operating cost (OPEX). When the rate of inflation increases, the cost of future repair will not be affected, but when it decreases, the present value of future repair will increase. In this chapter, we assume an inflation rate of about 4%, which depends on a country's general economy. Every country has its own published inflation rate.

9.2.2 Repair Time

The time required to repair the structure for steel corrosion is the time at which corrosion begins in the steel reinforcement bars and the time needed to spall the concrete cover with signs of concrete deterioration. This essentially requires work with repair. Tutti (1982) gave a simple explanation of the process of corrosion with time; the steps were for all types of corrosion, and there was no difference if corrosion happened as a result of chloride attack or carbonation propagation. However, the invasion of chlorides, as well as the carbonation of concrete, takes a long time to break the passive protection layer on the steel bars and start corrosion.

After that, from the beginning of corrosion to a significant deterioration in the concrete, when repair will be necessary, will take more time. Raupach (1996) pointed out that in the case of concrete bridges, degradation occurs in about 2–5 years. Therefore, the time for repair is the total time required for the protection of depassivation in addition to 3 years.

The steps of corrosion's effect on a concrete structure are illustrated in Figure 9.1. Note that after construction, it will take time until chloride concentration or carbonation accumulates on a structure's surface and then spreads into the concrete, as shown in the figure. The next step is the propagation of chlorides or carbonation until the steel bars are reached. The third step is the start of corrosion on the steel bar, which at this time will have an impact on the concrete strength, by reducing steel diameter and cracks occurring on the concrete surface. The last step is an increase in the crack width until spalling of the concrete cover.

From the preceding analysis, we find that the time required for repair depends on the time needed to increase the percentage of chloride concentration to the limit that will initiate the corrosion, in addition to the rate of corrosion, which will happen after that. The previous chapter discussed several methods for protecting a structure from corrosion. These different methods delay the start of corrosion for a longer period of time as they reduce the rate of the chloride or carbonation propagation in concrete. They will reduce the rate of corrosion after that as well. Note that the preceding analysis relies on the noninterference of the hair cracks on the concrete in the rates of spread of chlorides or carbonation within the concrete. It is assumed that the design was based on the absence of an increase in the cracks more than that permissible in codes (as in Chapter 5) and also that the concrete was produced based on a quality control procedure according to the code; thus, the presence of cracks is assumed to be within allowable limits.

The time required to start the structural repair depends on the nature of the surrounding weather and environmental factors that affect the beginning as well as the rate of corrosion. This time is determined by knowing the rate of corrosion and the

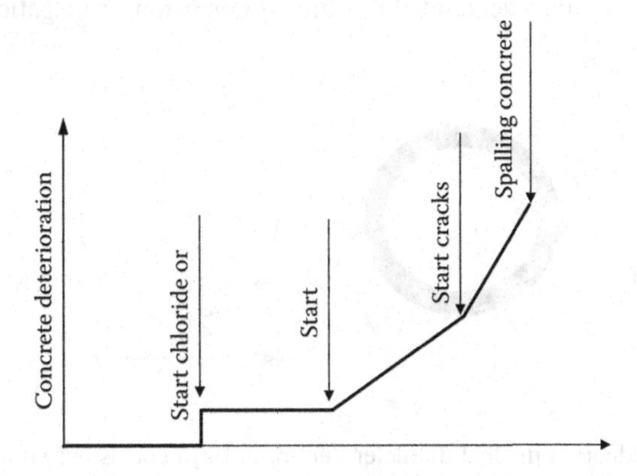

FIGURE 9.1 Sketch represents concrete structure deterioration process.

required time to spall the concrete cover. The deterioration of concrete increases the probability of a structure's collapse with time; some studies have identified the probability of failure, which should not be beyond the structural reliability classified in various specifications.

A study on residential buildings by El-Reedy and Ahmed (1998) focused on how to determine the appropriate time to perform repair following corrosion on concrete columns. It considered environmental conditions around the structures: They were affected by humidity and temperature, which have a large impact on the increasing rates of corrosion. A corrosion rate of 0.064 mm/year reflects dry air, while a rate of 0.114 mm/year is based on a very high moisture rate. This study took into account the increasing resistance of concrete over time as well as the method of determination in the case of higher steel or low steel columns, with different times required for the repair process.

9.2.3 Capacity Loss in Reinforced Concrete Sections

According to the fundamentals of design of any reinforced concrete member, the member's capacity depends on the cross-sectional dimensions (concrete and steel area) and material strength (concrete strength and steel yield strength). In the case of uniform corrosion, as shown in Figure 9.2, the total longitudinal reinforcement area can be expressed as a function of time t, as follows:

$$As(t) = \begin{cases} n\pi D^2/4 & \text{for } t \le T_i \\ n\pi\left[D - 2C_r(t - T_i)\right]^2/4 & \text{for } t > T_i \end{cases} \tag{9.2}$$

where

D is the diameter of the bar
n is the number of bars
T_i is the time of corrosion initiation
C_r is the rate of corrosion

This equation takes into account the uniform corrosion propagation process from all sides.

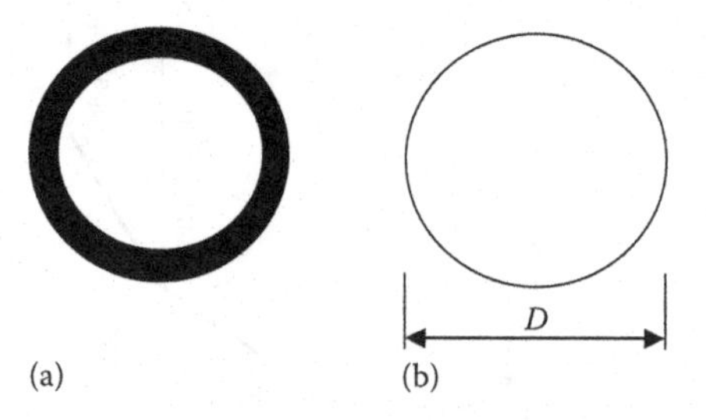

FIGURE 9.2 Reduction in steel diameter due to uniform corrosion: (a) uniform corrosion on the steel bars and (b) steel bars without corrosion.

The curves in Figures 9.3 and 9.4 show that, over time, the collapse of a structure is more likely with the increase in the rate of corrosion. Khalil et al. (2000) study mentioned earlier has revealed that concentrically loaded reinforced concrete columns will be due for maintenance about 4 or 5 years from the initial time of corrosion with the lowest steel ratio and higher corrosion rate. In cases of higher steel ratios and lower corrosion rates, this period may increase to about 15–20 years. Moreover, this study expects columns with eccentricity to increase the moment on

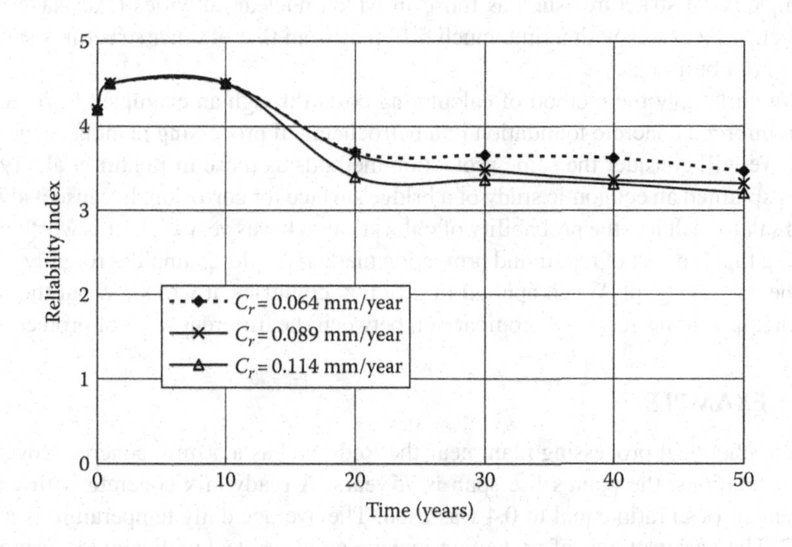

FIGURE 9.3 Effects of corrosion rate on the reliability index at reinforcement ratio equal to 1%.

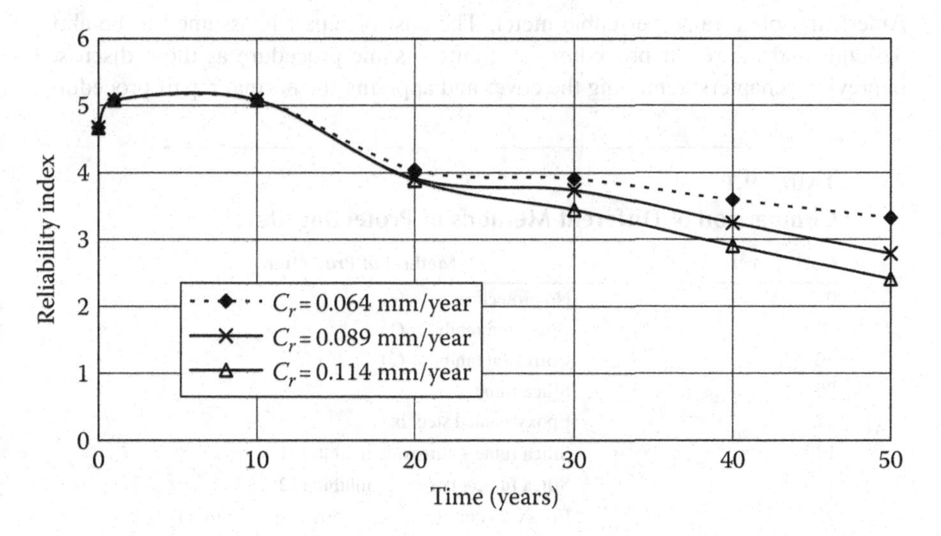

FIGURE 9.4 Effects of corrosion rate on the reliability index at reinforcement ratio equal to 4%.

the column, so the design should increase the percentage of steel. In such a case, it is expected that the deterioration of a structure and movement toward criticality will be after a couple of years in the case of the lowest reinforcing steel ratio and very high corrosion rate. On the other hand, in the case of higher reinforcing steel ratio and lower corrosion rate, the maintenance will be due after about 6 years.

As we expected, the time required for repair is linked closely to the nature of a structure and the method of design, as well as to the importance of the building. For example, vital structures such as those in which nuclear activities take place need protective measures with a time much different from that of other structures such as residential buildings.

We will apply the method of calculating costs through an example of protecting the reinforced concrete foundation in a petrochemical processing plant near the Red Sea. We will consider the same protection methods as those in Bentur et al. (1997), who explained an economic study of a bridge surface for corrosion; because it always used salt to melt ice, the probability of chloride attack was very high. It is worth mentioning that the cost of repair and protection methods in the example is roughly based on the cost in Egypt. When applied in different countries, it will vary, but the price is stated just to perform cost comparison between the different ways of protection.

9.3 EXAMPLE

A petrochemical processing plant near the Red Sea has a 50 mm concrete cover for all foundations; the plant's life span is 75 years. A ready-mix concrete with water-to-cement (*w*/*c*) ratio equal to 0.4 was used. The average daily temperature is about 26°C. The various types of protection include painting steel with epoxies, using silica fume, and using corrosion inhibitors. The initial cost estimate for protection by these methods is shown in Table 9.1; the value of cost is calculated at the rate of the American dollar value per cubic meter. The cost of repair is assumed to be about $\$200/m^3$ and the repair procedure is often the same procedure as those discussed in previous chapters: removing the cover and applying the normal repair procedure.

TABLE 9.1
Comparison of Different Methods of Protecting Steel

Cost ($/m³)	Method of Protection
0	No protection
40	Corrosion inhibitor (1)
50	Corrosion inhibitor (2)
80	Silica fume
55	Epoxy-coated steel bars
120	Silica fume + corrosion inhibitor (1)
130	Silica fume + corrosion inhibitor (2)
95	Epoxy-coated steel bars + corrosion inhibitor (1)
105	Epoxy-coated steel bars + corrosion inhibitor (2)
250	Cathodic protection system

From Table 9.1, corrosion inhibitor (1) is an anodic inhibitor such as calcium nitrate with a concentration of 10 kg/m^3 and corrosion inhibitor (2) is calcium nitrate with a concentration of 15 kg/m^3. The values of chloride ion content on the steel level are shown in this table. A discount rate of about 4% is imposed, as mentioned earlier, with the assumption that repair will be sufficient for 20 years only. All of these assumptions are taken into account when the comparison is made.

9.3.1 Required Time to Start of Corrosion

In the case of structures exposed to chloride, as in the previous example, the time required for the spread of chloride in the concrete until it reaches the steel is the time required for corrosion. Upon the chloride's arrival at a certain value limit, corrosion will start. The assumption prevalence rate of diffusion is about 1.63×10^{-12} m^2/s for concrete cast using a *w*/*c* ratio of about 0.4. It is assumed that the propagation rate is about 1.03×10^{-12} m^2/s when silica fume is used. The propagation rate and its impact are shown in Figure 9.1, where the time needed to increase the content of the chloride on a steel bar is a result of the previous propagation of chloride inside a 50 mm–thick concrete cover.

Berke et al. (1996) offered computational methods for time of the beginning of corrosion. The results of the calculations are shown in Table 9.2, which shows that the time needed to reach the amount of chloride to begin corrosion differs from the way for the protection of others. Note that there is no increase in the limits of chlorides that have occurred in the use of steel corrosion coated by epoxies or by using silica fume.

TABLE 9.2
Time Required for Start of Corrosion

Method of Protection	Chloride Limit (kg/m^3)	Time to Reach the Limit	First Repair Time	Second Repair Time	Third Repair Time
No protection	0.9	17	20	40	60
Corrosion inhibitor (1)	3.6	37	40	60	
Corrosion inhibitor (2)	5.9	+75			
Silica fume	0.9	22	25	45	65
Epoxy-coated steel bars	0.9	17	32	52	
Silica fume + corrosion inhibitor (1)	3.6	51	54		
Silica fume + corrosion inhibitor (2)	5.9	+75			
Epoxy-coated steel bars + corrosion inhibitor (1)	3.6	37	52		
Epoxy-coated steel bars + corrosion inhibitor (2)	5.9	+75			

In the case of structures exposed to carbonation spread inside the concrete cover, several equations can calculate the time required for this propagation (see Table 3.1 in Chapter 3). The following equation can deduce the depth of carbon transformation:

$$d = A(B)^{-0.5} \tag{9.3}$$

where A is a fixed amount depending on the permeability of the concrete as well as the quantity of carbon dioxide in the atmosphere and several other factors described in Chapter 3.

$$A = \left(17.04(w/c) - 6.52\right) \cdot S \cdot W \tag{9.4}$$

where

w/c is the ratio of water to cement (less than 0.6)
S is the effect of cement type
W is the weather effect

These equations give the average depth of the transformation of carbon. Therefore, when calculating the maximum depth of the transformation of carbon, one should increase from 5 to 10 mm. $S = 1.2$ in the case of using cement that has 60% slag. $W = 0.7$ in the case of concrete protected from the outside environment.

9.3.2 Time Required to Start of Deterioration

The time required after the beginning of corrosion has already been discussed. As stated before, this time is about 3 years in the absence of a corrosion inhibitor and can extend up to 4 years when a corrosion inhibitor is added. However, when epoxy-coated reinforcing steel is used, the period is extended to 15 years. Note that the use of reinforcing steel coated by epoxy helps reduce the rate of corrosion in a clear reversal of the contraceptive use of corrosion inhibitor as it is not actually affected in the reducing rate of corrosion.

Generally, the time of collapse of concrete cover from the beginning of corrosion depends on the rate of corrosion occurring in the steel reinforcement. Several studies have calculated the rate of corrosion. It was found that corrosion is closely related to relative humidity. As stated by Tutti (1982), in the case of steel corrosion due to carbonation, the corrosion rate decreases gradually at a relative humidity of 75% or less. The corrosion rate increases quickly when the relative humidity reaches 95%. It is noted that when the temperature decreases by about 10°, the corrosion rate will decrease by about 5%–10%.

Broomfield (1997) stated that the rate of corrosion is affected by the relative moisture, a dry situation in which concrete is poured, and the proportion of chlorides. Morinaga (1988) stated that the rate of corrosion was totally prevented when relative humidity was less than 45%, regardless of the chloride content and the temperature or oxygen concentration. Broomfield also stated that cracks in concrete occur when a lack of steel is about 0.1 mm and at less than 0.1 mm as well. That depends on the oxygen concentration and distribution, as well as the ability of concrete to withstand the excessive stresses. Sometimes, the placement of the steel bars causes cracks quickly, when bars are adjacent to each other or in the corner. In practical measurement, it is found that reduction of about 10–30 μm in a section results in a fragile layer of corrosion and is enough to cause concrete cracks.

The following equation of Kamal, Salama, and El-Abiary (1992) is used to calculate the time necessary from the beginning of corrosion for the emergence of the effects of corrosion on the concrete:

$$t_s = 0.08(C-5)/(D \cdot C_r) \tag{9.5}$$

where

t_s is the time from beginning of corrosion until the concrete cover falls
C is the concrete cover thickness
D is the steel bar diameter
C_r is the corrosion rate units (mm/year)

The total time expected to perform the first repair depends on the time of the beginning of corrosion in addition to the time needed to increase the rate of corrosion to cause concrete deterioration and to perform the repair.

9.3.3 Cost Analysis for Different Protection Methods

An equation was applied for the calculation of the present value of all methods of protection for steel reinforcement and is outlined in Table 9.2. The different values of the present value of each protection method have been clarified in Table 9.3, as have

TABLE 9.3
Cost Analysis for Different Protection Methods ($/m³)

Protection Method	Initial Cost	First Repair (NPV)	Second Repair (NPV)	Third Repair (NPV)	Total Repair Cost	Total Cost
No protection	0	91.3	41.65	19.03	151.98	151.98
Corrosion inhibitor (1)	40	41.66	19.03		60.69	64.69
Corrosion inhibitor (2)	50	0				50
Silica fume	80	75	42.22		117.22	197.22
Epoxy-coated steel bars	55	57.03	26		83.03	138.03
Silica fume + corrosion inhibitor (1)	120	24.05			24.05	144.05
Silica fume + corrosion inhibitor (2)	133	0			0	133
Epoxy-coated steel bars + corrosion inhibitor (1)	95	26			26	121
Epoxy-coated steel bars + corrosion inhibitor (2)	105	0				105

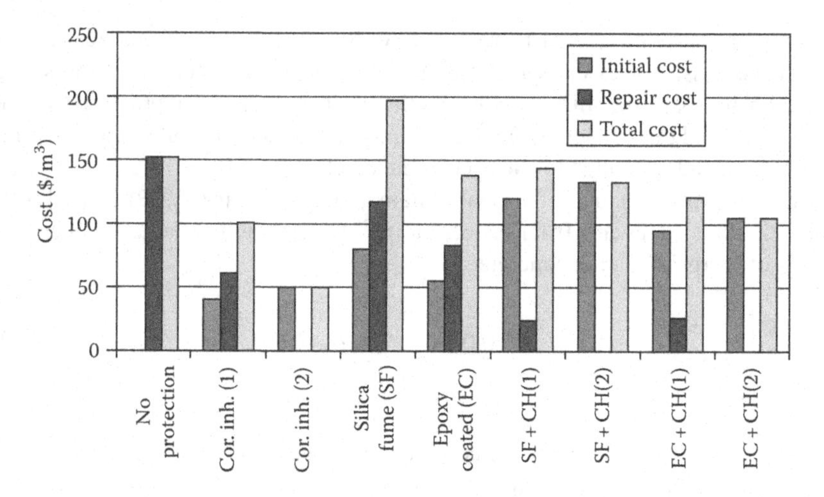

FIGURE 9.5 Economic comparison of different alternatives.

TABLE 9.4
Comparison of Initial Cost, Total Repair Cost, and Total Cost for Different Methods

Protection Method	Initial Cost	Total Repair Cost	Total Cost
No protection	0	151.98	151.98
Corrosion inhibitor (1)	40	60.69	100.69
Corrosion inhibitor (2)	50		50
Silica fume	80	117.22	197.22
Epoxy-coated steel bars	55	83.03	138.03
Silica fume + corrosion inhibitor (1)	120	24.05	144.05
Silica fume + corrosion inhibitor (2)	133	0	133
Epoxy-coated steel bars + corrosion inhibitor (1)	95	26	121
Epoxy-coated steel bars + corrosion inhibitor (2)	105		105

initial costs and the cost of repair. The total cost calculations of the current value are presented in Figure 9.5; three columns representing the cost of each method of protection are drawn in the graph.

Adding the cathode protection method costs about \$500–\$1500/m^3; even taking into account that the lower cost of the cathode protection is \$500/m^3, it will be much higher than the higher costs of other alternatives, as shown in the previous table. If the cathodic protection system is applied after 20 years, it would have a net present value (NPV) equal to around \$450 based on \$1000/m^3 cost and no operating, maintenance, or repair costs for the next 55 years.

In the comparison between Tables 9.3 and 9.4, one can choose the least costly method of protection and preferences. The previous example was a comparison among different methods of protection (Figure 9.5). However, the comparison must take into account the assumptions on which the calculations are based, including inflation and interest rates as well as labor and raw materials prices, which are recognized as initial costs.

9.4 REPAIR AND INSPECTION STRATEGY AND OPTIMIZATION

The decision-making methods of engineering projects have become the focus of numerous studies—as has their economic importance to the cost of the project as a whole—because a wrong decision could result in spending huge amounts of money to reverse it. Therefore, some studies have used decision trees to determine the appropriate time for repair, particularly in some bridges that need periodic inspection. When implementing a maintenance plan strategy based on cost, we must consider all the factors that affect corrosion (see Chapter 5). Any method upon which a decision is based should define the time for regular maintenance (e.g., every 10, 15, or 20 years). This decision is reached first and foremost through the identification of the least expensive option.

Figure 9.6 summarizes the reliability of any structure during its lifetime. As one can see, the structure in the beginning will have higher capacity; the probability of failure value depends on the code or standard on which the structure design is based. After a period of time, the structure will deteriorate, thus increasing its probability of failure. After time Δt, when the inspection and repair have been performed, the structure will recover its original strength, as shown in the figure. After another period of time, the probability of failure will increase to a certain limit and then more maintenance will be performed, and so on.

Inspection alone does not improve reliability unless it is accompanied by a corrective action when a defect is discovered. Some policies and strategies used in a wide range of concrete structures, which have been programmed, include the following:

- Monitoring until the crack depth reaches a certain proportion of the material thickness, and then repairing
- Immediately repairing on the detection of indications of damage
- Repairing at a fixed time (e.g., 1 year) after the detection of indications of damage
- Repairing as new (i.e., repair welding)

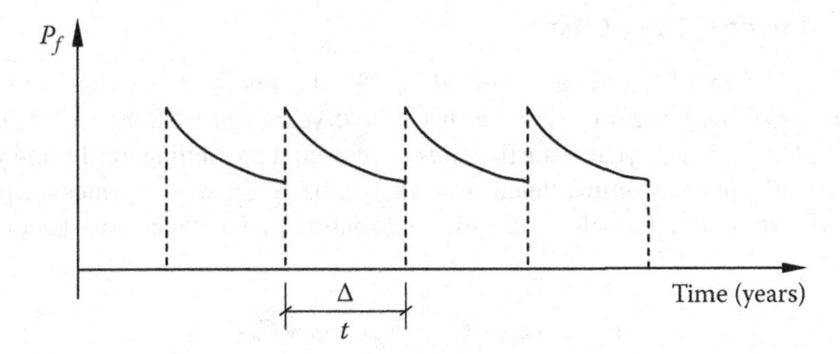

FIGURE 9.6 Concrete structure performance.

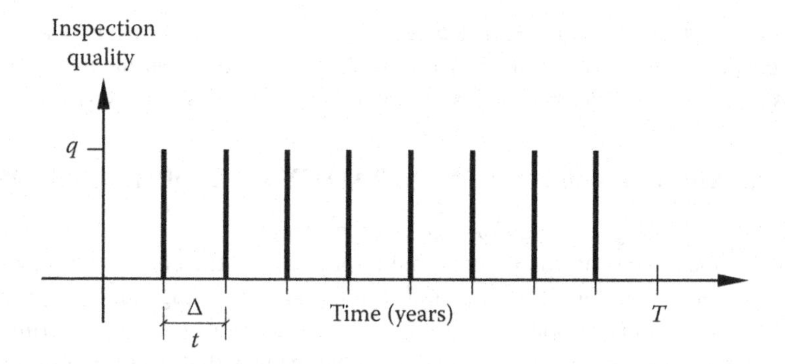

FIGURE 9.7 Inspection strategy.

Generally, it is assumed that inspections are performed at constant time intervals, as shown in Figure 9.7, since the inspection authorities often prefer a constant inspection interval to facilitate planning. From this point, the inspection intervals are chosen so that the expected cost of inspection, repair, and failure is minimized. While most concrete structure inspection techniques are visual or involve various nondestructive testing inspection methods, the ability to detect damage depends on the quality of the inspection performed. The higher the quality of the inspection method, the more dependable will be the assessment of damage. No repair will be made unless the damage is detected.

9.4.1 Repair

Inspection may not affect the probability of failure of a structure. Following an inspection in which damage is found, a decision must be made regarding repair. This decision will depend on the quality of the inspection. With advanced inspection methods, even a small defect can be detected and repaired. High-quality inspection may lead to high-quality repair, which brings the reliability of the structure closer to its original condition. Aging has an effect on the structure such that its reliability is decreased. This chapter proposes that after the inspection and the repair, the structure's capacity will be the same as that in the design conditions, which are clearly shown in Figure 9.4.

9.4.2 Expected Total Cost

As mentioned in El-Reedy and Ahmed (1998), the first step is to determine the service life of the structure. Assume that it is 75 years and routine maintenance is scheduled once every 2 years (starting at $t=2$ years and continuing until $t=74$ years). Consequently, preventive maintenance work will be performed 37 times during the life of the structure. Therefore, the lifetime routine maintenance cost becomes as follows:

$$C_{FM} = C_{m2} + C_{m4} + C_{m6} + \cdots + C_{m74} \tag{9.6}$$

where C_{FM} indicates the total maintenance cost, and the total expected cost in its lifetime (T) is based on the present value worth. The expected lifetime preventive maintenance cost becomes as follows:

$$C_{IR} = C_{IR2}\frac{1}{(1+r)^2} + C_{IR4}\frac{1}{(1+r)^4} + C_{IR6}\frac{1}{(1+r)^6} + \cdots + C_{IR74}\frac{1}{(1+r)^{74}} \quad (9.7)$$

where
C_{IR} is the periodic inspection and minor repair
r is the net discount rate of money

In general, for a strategy involving m lifetime inspections, the total expected inspection cost is

$$C_{ins} = \sum_{i=1}^{m} \square C_{ins} + C_R \frac{1}{(1+r)^{T1}} \quad (9.8)$$

where
C_{ins} is the inspection cost based on the inspection method
C_R is the repair cost
r is the net discount rate

Finally, the expected total cost, C_{ET}, is the sum of its components, including the initial cost of the structure and expected costs of routine maintenance, preventive maintenance (including inspection and repair maintenance costs), and failure. Accordingly, C_{ET} can be expressed as follows:

$$C_{ET} = C_T + (C_{ins} + C_R)(1 - P_f) + C_f \cdot P_f \quad (9.9)$$

The objective remains to develop a strategy that minimizes C_{ET} while keeping the lifetime reliability of the structure above a minimum allowable value.

9.4.3 Optimization Strategy

To implement an optimum lifetime strategy, the following problem must be solved:

$$\text{Minimize}\, C_{ET} \text{ subjected to } P_{flife} \le P_{max}$$

where $P_{\max}$ is the maximum acceptable lifetime failure probability. Alternatively, consider the reliability index:

$$\beta = \phi^{-1}(1 - P) \quad (9.10)$$

where ϕ is the standard normal distribution function. The optimum lifetime strategy is defined as the solution of the following mathematical problem:

$$\text{Minimize}\, C_{ET} \text{ subjected to } \beta_{life} \geq \beta_{min}$$

The optimal inspection strategy with regard to costs is determined by formulating an optimization problem. The objective function (C_{ET}) in this formulation is defined as including the periodic inspection cost, the minor joint repair cost, and the failure cost that includes the cost of major joint repair. The inspection periodic time (Δt) is the optimization variable constrained by the minimum index, β, specified by the code and the maximum periodic time.

The optimization problem may be mathematically written as follows: Find Δt, which minimizes the objective function:

$$C_{ET}(\Delta t) = (C_{IR})(1 - P_f(\Delta t))\left(\frac{(1+r)^T - 1}{\left((1+r)^{\Delta T} - 1\right)(1+r)^T}\right) + C_f P_f(\Delta t)\left(\frac{(1+r)^T - 1}{\left((1+r)^{\Delta T} - 1\right)^T (1+r)^T}\right) \quad (9.11)$$

subject to $\beta(t) \geq \beta^{min}$, $\Delta t \leq T$, where

C_{IR} is the periodic inspection and minor repair cost per inspection
C_f is the major repair cost
i is the real interest rate
β^{min} is the minimum acceptable reliability index
C_{IR} and C_f are assumed constant with time

Even though the failure cost is minimized (as part of the total cost), it is often necessary to constrain the reliability index to fulfill code requirements. Since T is the time period for the proposed repair, n is the total lifetime of the building, and C_{IR} is the cost of inspection and repairs, they are costs resulting from the collapse. The value of the expected cost at each period of time is a curve that follows that in Figure 9.8, defining the period of time required for the process of periodic maintenance, which is achieved less expensively, as in the curve that follows.

The period of time required for the process of the periodic maintenance, which is achieved at the highest cost, is Δt. The previous equation generally can be used for comparison between different types of repair with different time periods and different costs. It is calculated by adding the initial cost of the repair and the design curve. The decision can be determined in terms of the expected cost and risk of collapse and is calculated by using different methods to calculate the possibility of the collapse of the concrete member or the use of the approximate method as shown in Table 9.5. For the possibilities that are imposed at the beginning of the age of structure, the

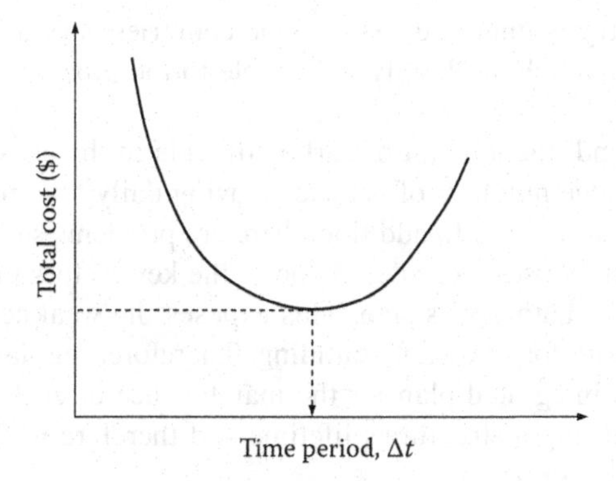

FIGURE 9.8 Optimization curve to obtain optimum maintenance time period.

TABLE 9.5
Relation between Structure Lifetime and Probability of Failure

Life (years)	0	10	20	30	40	50	60
Probability of failure (%)	0	9	25	50	75	91	100

probability of collapse at the age of zero is zero and the structure lifetime is assumed to be 60 years old. The possibility of the collapse is 100% at this age, and then the probability of failure at every period is calculated as shown in Table 9.5.

Some special software programs for the management of bridge systems (BMS), as identified by Abd El-Kader and Al-Kulaib (1998), define the time required to conduct the inspection and maintenance of each bridge. They consider the rate of deterioration in bridges with the lifetime of the structure. One such program includes a previous section that has been studied, a limitation on the time required for periodic inspection and maintenance work, and how to determine the cost, which depends on the budget planning for the repair of bridges. The program was developed by the Federal Highways Association.

The maintenance plan and its implementation depend on the criticality and the importunacy of the building. In order to proceed with maintenance, it is necessary to consider the sequences of failure. For example, concrete columns and concrete slab are different. Cost of repairing the two members may be the same or, in some situations, the repair of the concrete slab on the grade will cost more than repairing the columns. Which should be repaired first? Obviously, repair will start on the concrete columns because failure will have bad consequences from an economic point of view. Therefore, as mentioned earlier, the previous equation is a general equation and so all the factors that affect decision-making must be taken into consideration.

Generally, to proceed with maintenance on a regular basis in a certain defined time period is called preventive maintenance. It is important because it extends the structure's lifetime and maintains real estate that represents economic wealth for a country—especially in developing countries, in which all the material used in the construction industry is imported. Also, some countries have a huge investment in their coastal cities, which are highly vulnerable to the process of corrosion due to chloride effects.

On the other hand, the amount of carbon dioxide in the atmosphere in cities is increasing due to huge numbers of vehicles moving daily; this increases the corrosion of reinforcing steel bars. In addition, there are problems such as sanitation and drainage systems in houses, considered one of the key factors in corrosion of steel reinforcement of the bathroom's slab. This exposes the weakness of the structure and is a serious issue for the whole building. Therefore, regular maintenance and development of an integrated plan for the maintenance of each structure are very important in preserving a structure's lifetime and therefore to the national wealth of a country.

9.5 MAINTENANCE PLAN

In any organization, the maintenance team is responsible for maintaining the structural reliability of the buildings. The responsibility of this team lies clearly with the ministry of transportation, which is responsible for bridge maintenance. For example, in the United States, the famous Golden Gate Bridge is maintained by a team that regularly starts repainting the first section of the bridge as soon as it has finished painting the last section.

When we discussed the time spent performing maintenance earlier, we considered the probability of structural failure. From a practical point of view, it is complicated to calculate this accurately as it is needed for research rather than for practical issues. The calculation of the probability of failure and its consequences to obtain the structural risk is called the quantitative risk assessment. The main challenge is to calculate the structure probability of failure, which needs a special software and special reliability analysis method as per El-Reedy (2012). The other popular method is the qualitative risk assessment; it is easy and can be handled by the maintenance team without outside resources.

9.5.1 Assessment Process

The assessment of a concrete structure depends on its structural type, location, and existing load and operation requirements. For studying the risk assessment, different critical items that have an economic effect must be considered. The general definition of risk is summarized in the following equation:

$$\text{Risk} = \text{Probability of failure} \times \text{Consequences}$$

To define the risk assessment of any concrete structure, the factors that may affect the business's economics are worth considering. The structural risk assessment is represented by the probability of failure and its consequences; the whole building, part of the building, or only a concrete member may fail.

The first step to define and calculate the qualitative risk assessment is to hold a meeting with the maintenance team, and maybe with a consultant engineering company when huge structures are involved. Within this meeting, which will take several hours, different discussions define the factors that affect the probability of structural failure, such as corrosion of the steel reinforcement bars, overload compared to the design, new data revealing lower concrete strength than the design and project specifications called for, and the redundancy of the structure itself. The required data about the corrosion of steel bars, concrete compressive strength, and other factors have been discussed before and can be obtained by collecting the data and performing the visual inspection, as well as performing detailed inspection using measurement equipment such as ultrasonic pulse velocity, rebound hammer, or other available techniques. The factor that has the greatest influence on estimating the probability of structural failure is the structure's redundancy, which can be explained by Figures 9.9 and 9.10.

A beam needs to be designed, and the following two solutions are offered: (1) a structural system to be fixed at the two ends or (2) a structural system hinged at the two ends. Which system should be chosen? Take 5 minutes to think about this.

Many factors control a decision in selecting a suitable system. The following give the advantages and disadvantages for these two systems:

- Structural system 1:
  - The beam's cross section will be small.
  - The connections will be big and complicated as they have shearing force and moment.
  - It is reasonable from an architectural point of view.
  - The construction is complicated in connection.

- Structural system 2:
  - The beam cross section will be large.
  - The connection will be small as it is designed to shear force only.
  - It is easy to construct because the connection is simple.

After discussing these two systems, maybe the decision will be for in-house engineers and workers to carry out the construction. To avoid any problems, the simple beam option will be chosen. This type of decision is almost always made as the maintenance point of view is forgotten. Figure 9.10 shows the steps of collapse failure.

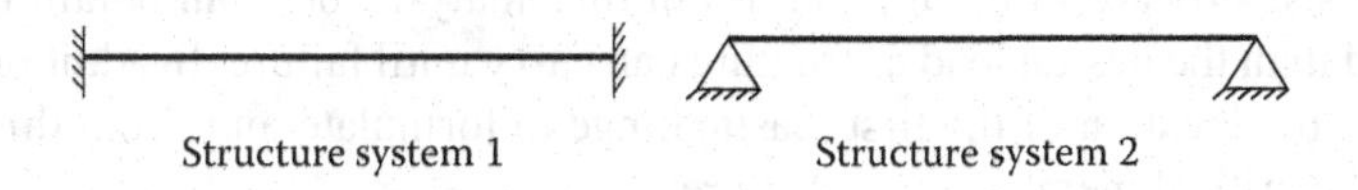

FIGURE 9.9 Comparison of structural redundancy.

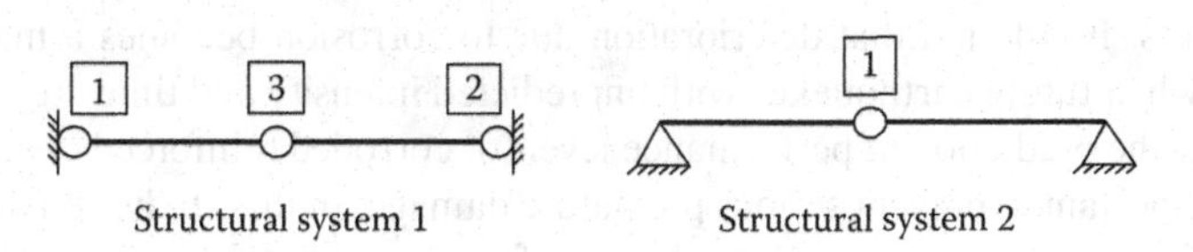

FIGURE 9.10 Comparison of structural redundancy.

Structural system 2, which is the simple beam, assumes that the load is increased gradually and the beam can accommodate the load until a plastic hinge is formed in the middle of the beam at the point of maximum bending moment; then, collapse will occur. On the other hand, for structural system 1 that is fixed at the two ends, when the load is increased gradually, the weaker of the left or right connection will fail first. As shown in the figure, the plastic hinge will form on the left connection and increase the load; the other connection will be a plastic hinge, so it is now working as a simple beam. By increasing the load, plastic hinge 3 will be formulated and then collapse will occur.

From the preceding discussion, one can find that structural system 1 will take more time to collapse because it fails after three stages. Structural system 2 fails in the first stage, so system 1 is more redundant than system 2. Moreover, when a comparison is made among different reinforced concrete members such as slabs, beams, columns, and cantilevers, one can learn that some members are more critical than others. A cantilever is most critical because any defect in it will have a high deflection and then failure. In addition, when a column fails, the load is distributed to other columns until the whole building fails. Thus, the column has low redundancy and is very critical because any failure on it will result in the failing of the whole building. However, the cantilever failure will be a member failure only, so we can go through the consequences from this approach.

In the case of beams and slabs, when a slab is designed, a simple beam is assumed and the maximum moment is calculated in the middle. Then, the concrete slab is designed by choosing the slab thickness and the steel reinforcement. The selected steel reinforcement will be distributed along the whole span. Theoretically, by increasing the load, the failure will be at a point, but actually the surrounding area will carry part of this load and so the redundancy of the slab is very high. Some research has mentioned that the reinforced concrete slab can accommodate loads that are twice the design load. On the other hand it is found by Zhang and Li in 2022 that in case of the structure under seismic load that corrosion of reinforcements had a substantial effect on the strength and lateral drift capacity of the joints between beams and columns. These two factors of corrosion and seismic should be considered in the model of likelihood of failure.

For the structure as a whole, we can use the pushover analysis to obtain the redundancy of the structure. This analysis is nonlinear and is now available for any structure analysis software in the market. From this analysis, one can obtain how much more load than the design load a structure can carry until failure. In addition, one can determine the location of the first plastic hinge to formulate and, from this, one can find out the critical member in a structure.

There was a study done by Yalciner et al. (2015) to apply the effect of corrosion and the seismic load on the building it will be done for a case study for school with age 15 years. It is found that deterioration due to corrosion becomes a more serious problem when future earthquakes with unpredicted intensity and time are considered. Therefore, the prediction of performance levels of corroded reinforced concrete structures is important to prevent serious premature damage. In this study, considering corrosion effects and the relevant data obtained from the structure were used to predict its performance level for different time periods by combining two major effects of corrosion. Deformation due to bond-slip relationships and loss in cross-sectional areas of reinforcement bars were examined as a function of corrosion rate for five corrosion levels. Plastic hinges were defined as a consequence of corrosion effects, and they were used to perform nonlinear pushover analyses. Incremental dynamic analysis was then performed for 20 individual earthquake ground motion records to predict the structure's time-dependent performance levels as a function of corrosion rate. Results of this study showed that corrosion had a serious effect on the performance levels of the school building considered in this study by decreasing bond strength.

Therefore, it is necessary to put a maintenance plan in place for a concrete structure. The first step is to discuss all the structural parts that need to be inspected and repaired and capture this information in a table. The team will put all the factors that affect the probability of structural failure into this table. Different factors will be assigned different values. For example, in the case of a redundancy factor, values will run from 1 to 10, and a column will have a value of 10 and a slab will have a value of 2. All the members of the team must agree on these numbers, based on their experience. The same values or higher values will be repeated for other factors, such as age, the engineering office that produced the design, the contractor who handled the construction, and the code used in design. As discussed earlier, in some countries, those responsible for construction have used seawater in mixing water, and some codes in the past agreed to use 6mm diameter steel in stirrups. Therefore, if corrosion occurred, these stirrups could not be seen because complete corrosion would have taken place.

Table 9.6 contains the probability of structural failure factors, but when this table is used, it must match with structural requirements. The table is not limited to these factors only; it is a simple example of calculating the quantity of the probability of failure. However, when a meeting is held, it is necessary to develop criteria about the number. For example, consider an engineering office that produces the design. If it is professional and well known in the market and one has had no problems with the firm, then lower values should be assigned. For an incompetent firm with which one has had problems in the same type of structure, a higher value is assigned. This procedure will apply also to the contractor.

The code value comes from the team's experience in considering problems not only for its own buildings but also for buildings throughout the country. Therefore,

TABLE 9.6
Factors' Effects on Probability of Structure Failure

Structure	Redundancy (1–10)	Age (1–10)	Designer (1–10)	Contractor (1–10)	Code (1–10)	Last Inspection (1–10)	Total Score
S1	8	10	6	6	9	10	49
S2	5	2	7	3	3	10	30
S3	3	5	8	10	5	2	33
S4	7	4	2	5	5	1	24
S5	1	3	4	2	5	10	25
S6	3	1	4	5	6	7	26

in some complicated structures or huge facilities, it is worth contacting a competent engineering office for a maintenance plan.

If one conducts no inspections, the building is a 'black box' and should be given the highest value. The things that one knows are bad are less critical than things about which one knows nothing. Then, it is necessary to go through the consequences with the same approach, considering the economic impact. Economic impact is the answer to the question of what the effect will be from an economic point of view if the whole structure or part of it fails (see Table 9.7).

There are many hazard factors that will guide structure risk assessment. The very first factors are the location itself as well as its expenses. Moreover, if it carries a processing facility for industry, then that must be taken into consideration because the failure may cause hazards or stop production. For example, consider the foundation under machines; any failure in it will influence the performance of the machine and will be the main cause for the processing plant's shutting down and stopping production. We calculate the weight of the consequences by the same procedure as that for probability of failure. Now that we have the weight of the probability of failure and the consequences, we calculate the risk assessment from Table 9.8.

TABLE 9.7
Consequence Weight

Structure	Impact on Person (1–10)	Impact on Cost (1–10)	Impact on Environment (1–10)	Impact on Repetition (1–10)	Total Score
S1	8	10	6	6	30
S2	5	2	7	3	17
S3	3	5	8	10	26
S4	7	4	2	5	18
S5	1	3	4	2	10
S6	4	5	3	5	17

TABLE 9.8
Risk Weights for All Structures

Structure	Probability of Failure	Consequences	Risk
S1	49	30	1470
S2	30	17	510
S3	33	26	858
S4	24	18	432
S5	25	10	250
S6	26	17	442

9.5.2 Risk-Based Inspection Maintenance Plan

After we calculate the risk assessment as shown in the previous Table 9.8, we classify the maintenance plan. The top third of the top risk structures will be red, the second third will be yellow, and the remaining third will be green; the priority list will be as shown in Table 9.9. From this table, it can be seen that the structures S1 and S3 are considered the critical structures upon which inspection would start. Structures S2 and S6, as the second priority, would be inspected after the first two structures. The last structures, S4 and S5, will be at less risk, so inspection would not be needed at this time.

As shown in the table, this simplified method can be used to plan maintenance for inspection and repair, taking into account that the budget is a very important factor in the maintenance plan. From the previous example, the budget may be enough this year to inspect structure S1 only, so it will be necessary to plan for the following years—maybe for the next 5 years. After the inspection is performed, the data will be analyzed through the system as in the flow chart in Figure 9.11.

TABLE 9.9
Structure Priority List

Structure Priority	Color Code	Risk Value
S1	Red	1470
S3	Red	858
S2	Yellow	510
S6	Yellow	442
S4	Green	432

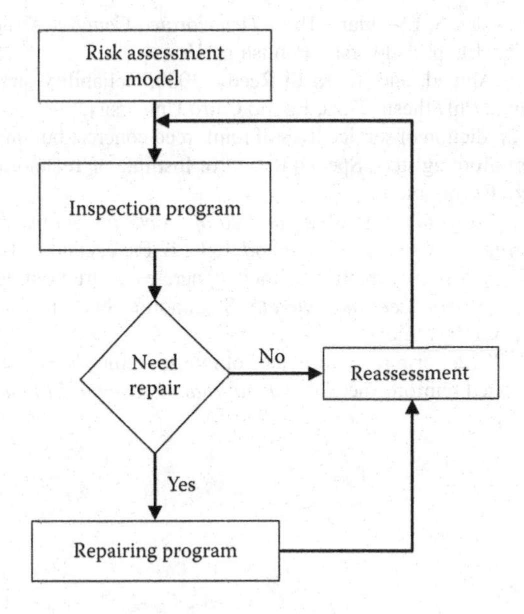

FIGURE 9.11 Risk assessment system.

Nowadays, for international organizations with a number of different buildings and structural elements, the integrity management system is very important for maintaining structures along their lifetimes and accumulating historical data. The closed loop of a structure risk assessment, an inspection program, a repair program, and reassessment is crucial to maintaining the structure in a good condition and as a reliable match with the operations requirements. The total quality control management system is very important for reviewing design, construction, and maintenance to be matched with operational requirements in cases of proposed development of or extension to a building or increasing its load. Therefore, it must be considered by a system of change management procedures to upgrade the previous risk-ranking table.

REFERENCES

Abd El-Kader, O. and A. Al-Kulaib. 1998. Kuwait bridge management system. *Eighth International Colloquium on Structural and Geotechnical Engineering*, Kuwait.

Bentur, A., S. Diamond, and N. S. Berke. 1997. *Steel Corrosion in Concrete*. London, UK: E & FN Spon.

Berke, N. S., M. C. Dallaire, M. C. Hicks, and A. C. McDonald. 1996. Holistic approach to durability of steel reinforced concrete. In *Concrete in the Service of Mankind: Radical Concrete Technology*, eds. R. K. Dhir and P. C. Hewelet. London, UK: E & FN Spon.

Broomfield, J. P. 1997. *Corrosion of Steel in Concrete*. London, UK: E & FN Spon.

El-Reedy, M. A. 2012. *Reinforced Concrete Structural Reliability*. Boca Raton, FL: CRC Press.

weEl-Reedy, M. A. and M. A. Ahmed. 1998. Reliability-based tubular joint of offshore structure-based inspection strategy. Paper presented at Offshore Mediterranean Conference (OMC), Society of Petroleum Engineers, Ravenna, Italy.

Yalciner, H., Sensoy, S., and Eren, O.. 2015. Seismic performance assessment of a corroded 50-year-old reinforced concrete building. *ASCE Journal of Structural Engineering 141*(12).

Kamal, S., O. Salama, and S. El-Abiary. 1992. *Deteriorated Concrete Structure and Methods of Repair*. Cairo, Egypt: University Publishing House.

Khalil, A. B., M. M. Ahmed, and M. A. El-Reedy. 2000. Reliability analysis of reinforced concrete column. PhD thesis, Giza, Egypt: Cairo University.

Morinaga, S. 1988. Prediction of service lives of reinforced concrete buildings based on rate of corrosion of reinforcing steel. Special Report of Institute of Technology, No. 23. Tokyo, Japan: Shimizu Corporation.

Raupach, M. (1996). *Corrosion of steel in the area of cracks in concrete laboratory test and calculation using transmission line method.* In C. L. Page, et al. (Eds.) 4th International symposium on corrosion of reinforcement in concrete construction, July 1–4, 1996, UK.

Tutti, K. 1982. *Corrosion of Steel in Concrete*. Stockholm, Sweden: Swedish Cement and Concrete Research Institute.

Zhang, X. and Li, B. 2022. Seismic performance of interior reinforced concrete beam–column joint with corroded reinforcement. *ASCE Journal of Structural Engineering 148*(2).

Index

Note: **Bold** page numbers refer to tables and *italic* page numbers refer to figures.

Printed in the United States
by Baker & Taylor Publisher Services

Printed in the United States
by Baker & Taylor Publisher Services